Essentials of Sports Nutrition

Second Edition

Essentials of Sports Nutrition

Second Edition

FRED BROUNS PhD
Nutrition and Toxicology Research Institute, Maastricht University, Maastricht, The Netherlands

CERESTAR–CARGILL
Vilvoorde R&D Center, Vilvoorde, Belgium

JOHN WILEY & SONS, LTD

First published 1993, as *Nutritional Needs of Athletes*
Copyright © 1993, 2002   John Wiley & Sons Ltd, The Atrium, Southern Gate, Chichester,
West Sussex PO19 8SQ, England

Telephone (+44) 1243 779777

Email (for orders and customer service enquiries) cs-books@wiley.co.uk
Visit our Home Page on www.wileyeurope.com or www.wiley.com

Other Wiley Editorial Offices

John Wiley & Sons Inc., 111 River Street, Hoboken, NJ 07030, USA

Jossey-Bass, 989 Market Street, San Francisco, CA 94103-1741, USA

Wiley-VCH Verlag GmbH, Boschstr. 12, D-69469 Weinheim, Germany

John Wiley & Sons Australia Ltd, 33 Park Road, Milton, Queensland 4064, Australia

John Wiley & Sons (Asia) Pte Ltd, 2 Clementi Loop #02-01, Jin Xing Distripark, Singapore
129809

John Wiley & Sons Canada Ltd, 22 Worcester Road, Etobicoke, Ontario, Canada M9W 1L1

Library of Congress Cataloging-in-Publication Data

Brouns, F. (Fred)
    Essentials of sports nutrition / Fred Brouns.—2nd ed.
      p. cm.
    New ed of: Nutritional needs of athletes / Fred Brouns. c1993.
    Includes bibliographical references and index.
    ISBN 0-471-49764-9 (cased)—ISBN 0-471-49765-7 (paper)
    1. Athletes—Nutrition. I. Brouns, F. (Fred). Nutritional needs of athletes. II. Title.

    TX361.A8 B74 2001
    613.2'024'796—dc21

                                                        2001026860

British Library Cataloguing in Publication Data

A catalogue record for this book is available from the British Library

ISBN 0-471-49764-9 (cased)
ISBN 0-471-49765-7 (paper)

Typeset in 10/12 Palatino by Mathematical Composition Setters Ltd, Salisbury, Wiltshire

This book is printed on acid-free paper responsibly manufactured from sustainable forestry
in which at least two trees are planted for each one used for paper production.

Contents

Preface

This book aims to give a scientific but easily understood overview of aspects related to nutrition and physical activity, especially of people involved in regular training with the goal to improve intensive sports performance.

The book refers to a large number of scientific reviews and publications, which have appeared in peer reviewed scientific journals. This means that these publications have survived the criticisms of the reviewers and that the interpretations are in line with existing scientific consensus.

To achieve a maximal degree of scientific consensus, the first draft of the manuscript was sent to a number of experts in the field of exercise and nutritional sciences. Selection of these experts was based on their actual research activities and their internationally known expertise in different fields of sport nutrition. Their reviews and criticisms were gratefully acknowledged and resulted in the final manuscript, published in 1993. This text received international attention, resulting in publication in German, French, Spanish and Japanese. The book's wide use as an educational text in graduate courses of sports sciences, physical education and sports medicine has led to numerous suggestions on how to further improve the contents. The current revised and updated book is the result of this process.

Fred Brouns

Acknowledgements

The helpful contributions of the following experts who critically reviewed and discussed the First Edition, to realize a status of scientific consensus in the final manuscript, have been gratefully acknowledged:

Prof. M. Williams	USA
Prof. W.H.M. Saris	Netherlands
Prof. Abel Mariné-Font	Spain
Prof. Dr. Clyde Williams	England
Prof. Ron J. Maughan	Scotland
Prof. Sigmund B. Strömme	Norway
Doz. Dr Peter Baumgartl	Austria
Prof. Michael Hamm	Germany
Dr Klaus-Jürgen Moch	Germany
Prof. Michel Rieu	France
Dr Charles-Yannick Guezennec	France
Dr Nancy J. Rehrer	New Zealand

1 Introduction

One of the most important nutritional aspects concerning athletes, recognized since the first competitions in ancient Greece, is the increased need for energy. Athletes involved in heavy physical activity need more food than more sedentary, less active people. The energy expenditure of a sedentary adult female/male amounts to approximately 1800–2800 kcal/-day. Physical activity by means of training or competition will increase the daily energy expenditure by 500 to >1000 kcal/h, depending on physical fitness, duration, type and intensity of sport. For this reason, athletes must adapt their energy intake by increased food consumption, according to the level of daily energy expenditure, in order to meet energy needs. This increased food intake should be well balanced with respect to the macronutrients (carbohydrate, fat and protein) and micronutrients (vitamins, minerals and trace elements). However, this is not always easy. Many athletic events are characterized by extremely high exercise intensities. As a result, energy expenditure over a short time period may be extremely high. Running a marathon, for example, costs about 2500–3000 kcal (137). Depending on the time needed to finish this may induce an energy expenditure of approximately 750 kcal/h in a recreational athlete and 1500 kcal/h in the elite athlete who finishes in approximately 2 to 2.5 h. A professional cycling race, such as the 'Tour de France' costs about 6500 kcal/day, a figure which will be increased to approximately 9000 kcal/day when cycling over a mountain pass (165).

Compensating for such high energy expenditure by ingesting normal solid meals will pose a problem to any athlete involved in such competitions, since digestion and absorption processes will be impaired during intensive physical activity. These problems are not restricted to competition days. During intensive training days, energy expenditure is also high (24). In such circumstances athletes tend to ingest a large number of 'in between meals', often composed of energy rich snacks, which, however, are often low in protein and micronutrients. As such, the diet often becomes imbalanced. Foods and drinks that are easily digestible and rapidly absorbed may solve this problem (23, 30). During endurance sports activity the body will also use its own energy stores (fat stored as adipose tissue and carbohydrate (CHO) stored as glycogen in liver and muscle). In addition, small amounts of (functional) protein (in the liver, gastrointestinal tract and muscle) will be broken down due to mechanical and metabolic stresses. These losses have to be compensated by supply of the necessary nutrients. At the same time heat will be produced which to a large extent

Figure 1 In professional cycling energy expenditure may exceed 9000 kcal (37 mJ) per day when cycling in the Alps

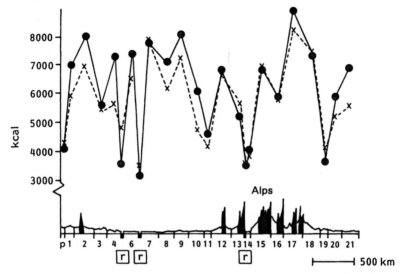

Figure 2 Daily energy expenditure (●—●) and energy intake (×— —×) as measured in a cyclist during the Tour de France. Note the extremely high energy expenditure as well as the ability to match energy balance with the use of liquid nutrition in addition to the normal meals. p = stage; r = rest day. Reproduced with permission from W. Saris *et al.* (165)

will be eliminated by the secretion and evaporation of sweat. As a result, fluid and electrolytes will be lost.

Large sweat losses may pose a risk to health by inducing severe dehydration, impaired blood circulation and heat transfer, leading to heat exhaustion and collapse (129, 143, 166, 167, 168). Insufficient replacement of CHO may lead to hypoglycaemia, central fatigue and exhaustion (16, 39, 43, 44, 113, 136, 137, 173, 174, 189). Inadequate protein intake induces protein loss, especially from muscle, and consequently a negative nitrogen balance and reduced performance (106, 107, 125).

These observations show that the requirements for nutrients and fluids should be met according to the level of daily physical activity and exercise. The type, intensity and duration of exercise will determine which nutritional measures and dietary interventions can be taken, particularly in the phases of preparation, competition and recuperation.

Problems related to increased needs for food and drink as well as the nutritional measures to solve them, concern not only the highly trained top athlete but also less trained sports people. This is especially the case when the latter, at equal absolute workloads, are prone to more stress, sweat more profusely, use more CHO as fuel for muscle work, utilize/break down more protein and recover more slowly from it. Highly trained individuals will work more economically and spend less energy to attain a certain mechanical work output than untrained subjects. Thus, anyone who tries to achieve a personal best and exercises at the upper level of his or her functional capacities will induce a maximal metabolic demand in order to meet the energy needs best with the available organ capacity. This is the case for an olympic athlete but also for a leisure athlete who is doing his utmost to complete a marathon. Well trained athletes have developed a larger metabolic capacity and, accordingly, have a better ability to run faster and to recover more quickly. However, when exercising at maximal speed (capacity), the trained athlete will also become energy depleted, dehydrated and exhausted at a certain moment. Therefore, from a qualitative point of view, training and exercise guidelines including nutritional measures are the same for professional and amateur athletes. Food and meals to be ingested shortly before and during exercise, or during a small break between exercise periods, should be adapted to specific ingestion and assimilation conditions, which depend on the nature and circumstances of the sport practised. For example, nutritional requirements are totally different for athletes competing in cycling, running or swimming events.

Some groups of athletes compete in sport events where a low body weight is necessary to perform well or to compete in a certain weight category. These athletes are on the one hand training frequently and intensively and spend a lot of energy, but, on the other hand they have to be careful with the intake of energy rich foods because they have to maintain a low body weight. The low energy intakes may in these situations lead to a marginal supply of essential nutrients such as protein, iron, calcium, zinc, magnesium and vitamins. The need for a relatively high carbohydrate intake, to balance the carbohydrate used in muscle work may also be compromised. This aspect should receive special attention as many of these athletes are young and still in a period of growth and development (163).

Depending on the type of sport and training it is possible to categorize athletes at risk for marginal nutrient intake. These athletes and those who combine heavy training with weight reduction programmes should be counselled most intensively in order to benefit from nutritional measures to optimize the diet (Table 1). The ability to take safe nutritional measures depends on the availability of standards, well documented guidelines and appropriate legal regulations for this category of products. Fitness and health focused people should be informed by health/sports professionals about the role of diet in their sport (180).

Especially in endurance events athletes should be able to make an appropriate choice from available food items, including food products designed and marketed for them. The nutritional education of athletes and their coaches warrants attention in this respect. Several studies have shown that their nutritional knowledge is marginal, despite the fact that awareness about the importance of nutrition is growing and articles about nutrition are regularly published in athletic journals (46, 149, 198). Legal food regulations

Table 1 High risk sports for marginal nutrition

Criteria	Sports discipline
Low weight—chronically low energy intakes to achieve low body fat	Gymnastics, jockeys, ballet, dancing, rhythmic gymnastics, ice dancing, aerobics
Competition weight—drastic weight loss regimens to achieve desired weight category	Weight class sports (e.g. judo, boxing, wrestling, rowing, ski jumping)
Low fat—drastic weight loss to achieve lowest possible body fat	Body building
Vegetarian athletes	Especially in endurance events

exist for foods covering special needs in special circumstances, e.g. dietetic food products for ill people or other subjects with special physiological conditions. However, no such regulation is available (yet) for sport foods and sport supplements. The aim of a food product regulation is to lay down scientifically acceptable conditions to ensure that food products are of an acceptable quality standard. Also, the (benefit) claims made for the product should have a scientific basis that is generally accepted as valid. This should also apply to food products that are labelled as sports food. One general problem in this respect is that the current recommended dietary intakes, which are differentiated for age, sex and daily activity level, may not be appropriate for the athlete for a number of nutrients. The aim of this book is to systematically describe the nutritional aspects of sports. The chapters presented may form a practical basis for anyone who wants to be informed about the essentials of sports nutrition. Recently, major scientific reviews on these topics have been published. One complete review was the output of an international scientific consensus meeting on nutrition and sport, held in the IOC headquarters (March 1991, Lausanne, Switzerland, proceedings published in a special issue of the *Journal of Sport Sciences*, Volume 9, summer 1991). The development in knowledge, based on sound research over the last 10 years, has been tremendous. The references are listed at the end of the book and supply the interested reader with more details on the topics described.

Key points

- In athletes, an adequate intake of nutrients is essential for the maintenance of an appropriate nutritional status, optimal performance and recovery as well as the reduction of health risks associated with regular highly intensive exercise.

- A very large energy turnover, for example as takes place during intensive endurance events, requires an adequate energy and nutrient intake in order to maintain energy, nitrogen and fluid balance.
- The large gastrointestinal bulk associated with a carbohydrate rich diet, when consuming normal food, causes many endurance athletes to change their food habits from three main meals to more frequent smaller meals and to ingest 30–50% of daily energy intake as in-between meals/snacks. These are often high in energy but low in dietary fibre, protein and micronutrients. This may lead to a reduction in the quality of the diet in terms of nutrient density.
- Nutritional education of athletes and coaches is important. However, although many nutrition orientated publications appear in athletic journals, the evidence shows that nutritional knowledge among athletes and their significant others needs to be improved (see also Parts I and III).
- Athletes who ingest chronically or repeatedly low energy diets, such as gymnasts, dancers, low weight class athletes, bodybuilders and female distance runners, are at potential risk of developing nutritional disorders and poor nutritional status. Such athletes can improve their nutrient intake and nutrition status by choosing foods, food products and food supplements that have an enhanced nutrient density for specific nutrients. Appropriate nutritional advice for both athlete and coach/parents is essential in this respect.

I Nutritional Aspects of Macronutrients in Sport

2   Carbohydrate (CHO)

CHO is the most important fuel for high intensity muscular work. To demonstrate the importance of CHO for performance and recovery, this chapter briefly describes how CHO makes up part of the energy reserves in our body and how CHO metabolism is influenced by exercise (see also Chapter 13).

CARBOHYDRATE RESERVES

In the body CHO is stored as long chains of glucose units, called glycogen, in the liver and in the muscles. This form of storage is in principle comparable to that of starch present in potatoes, banana and other plant foods.

LIVER GLYCOGEN

The amount of glycogen stored in the liver amounts to approximately 100 g. This quantity may change periodically depending on the amount of glycogen that is broken down for the supply of blood glucose in periods of fasting and the amount of glucose that is supplied to the liver after food intake. Accordingly, liver glycogen reserves increase after meals but diminish in between, especially during the night, when the liver constantly delivers glucose into the bloodstream to maintain a normal blood glucose level (89–91, 135, 158). A constant blood glucose level, within a narrow physiological range, is important because blood glucose is the primary energy source for the nervous system.

Influence of Exercise

During physical exercise a number of metabolic and hormonal stimuli will lead to an increased uptake of blood glucose by the working muscles to serve as a fuel for muscular contractions. To avoid the blood glucose level falling below the normal physiological value, the liver will at the same time be stimulated to supply glucose to the bloodstream. This supply is mainly derived from the liver glycogen pool and to a small degree from the process of gluconeogenesis (*de novo* glucose synthesis) by the liver cells from precursors such as amino acids (1, 90, 135, 158). Thus, appropriate glycogen availability in the liver is a key factor for maintenance of a normal blood

Figure 3 Recreational runners performing long distance competitions and trying to set a new personal best perform top sport

glucose level during prolonged exercise. As soon as the liver glycogen store is emptied and exercise is executed without concomitant food intake, the liver may become glycogen depleted. Since the blood glucose utilization and uptake during exercise by active muscles remains high, blood glucose may than fall to hypoglycaemic levels. Glucose uptake by the muscles from blood will drop to marginal levels and the working muscles will then totally depend on the local CHO supply from remaining muscle glycogen. Depending on the rate at which hypoglycaemia develops, this may or may not impair performance capacity. Central as well as local fatigue may then occur. This phenomenon has been well described both in sports practice and in scientific studies (39, 44, 63, 65, 91, 173, 174).

A condition of hypoglycaemia during exercise will gradually induce the maximal use of alternative fuels such as fat and protein and therefore stimulate fat mobilization, protein breakdown and the use of fatty acids and amino acids. The best way to circumvent the consequences of a developing CHO shortage during exercise is to maintain an appropriate CHO supply to the blood by means of oral intake.

MUSCLE GLYCOGEN

The amount of glycogen that is stored in total muscle in the body amounts to approximately 300 g in sedentary people and may be increased to >500 g in trained individuals by a combination of exercise and the consumption of a CHO rich diet (16, 90, 174). The total intramuscular stored CHO may thus range in energetic equivalent from 1200 to 2000 kcal.

Influence of Exercise

The rate at which muscle glycogen is mobilized for the production of energy needed for muscle contraction depends on the training status of the athlete as well as on the duration and intensity of the exercise.

Research has shown that a very small pool of energy rich phosphates (adenosine triphosphate and creatine phosphate), which is immediately available for muscle contractions at any moment of suddenly increased energy need, may deliver energy for a period of up to maximally 10–15 s. For any longer lasting events, the energy requirements for muscle work will have to be covered by the mobilization and subsequent metabolism of substrates from the CHO and fat pools in muscle, liver and adipose tissue (15, 16, 19, 90, 136, 137, 173). The use of any of these pools will never be exclusive. Thus, at any time muscle will use a mixture of CHO, fat and (to a very small degree) protein/amino acids for energy production.

However, depending on exercise intensity and duration, one of the fuels may become the major energy deliverant. For example, at rest practically all

Figure 4   Individuals engaged in sport and ingesting low energetic diets are prone to marginal or insufficient intake of key nutrients

Figure 5 Swiss national coaches learn about food and cooking in special courses on nutrition and top sport

the energy needed for the resting metabolism is derived from fat, with the exception of the central nervous system and the red blood cells, which rely primarily on blood glucose. In this situation the possible energy supply ratio may be in the order of 90% from fat and about 10% from CHO. During a situation of increased physical activity, i.e. light physical work, or a moderately intensive sports activity, the body will use metabolic, hormonal and nervous control mechanisms to mobilize glucose from glycogen pools to serve as a rapid energy deliverant (133). Synchronously the mobilization of fatty acids will be stimulated. After about 20–30 min a new metabolic steady state will be achieved in which the energy supply ratio of fat to CHO may be about 50% : 50%. Thus, a gradual shift from high fat/low CHO utilization at rest to enhanced CHO utilization during physical activity has taken place.

At even higher work intensities the body will start to use more and more CHO as fuel. Accordingly, during strenuous sports activity, such as middle distance running, 1000 m speed skating and other events lasting 1–3 min, CHO will become the most important fuel (13, 39, 44, 79, 174). The ratio of fat to CHO may then be 10% : 90%. The reason for this shift to the dominant use of CHO is that the maximal amount of energy that can be produced from CHO, per unit of time, is higher than that of fat. In addition, the amount of oxygen required for energy production from CHO is about 10% lower than that of fat (119). Besides these energetic advantages the process of substrate mobilization, substrate transport to and uptake by muscle cells also plays a role. This process is relatively fast in the case of CHO and slow with fat. Thus, it turns out that the muscle shifts to the most economical and rapidly available energy source in periods of suddenly increased energy requirements. Such a fine-tuned regulation of energy rich substrate selection enables athletes to work at a higher intensity when using CHO as the main energy source.

Indeed, several lines of evidence show that intense and lasting muscle work cannot be performed without appropriate availability of CHO. As soon as specific muscles or muscle fibres become glycogen depleted they will be impaired in their ability to perform repeated high intensity contractions (16, 39, 90, 113). Research shows that glycogen depletion, either by exercise or by a combination of exercise and low CHO intake, leads to a reduction in work capacity of about 50% of the normal maximal working capacity (39, 113, 137). Alternatively, when the CHO stores in muscle and liver are increased by diet manipulation, athletes are able to perform longer at high exercise intensity. These examples show that the availability of CHO and the size of the glycogen stores are important and limiting factors for endurance performance.

Moreover, CHO is also the prime energy source for the central nervous system and for the red blood cells. CHO is also required for the deliverance of pyruvate to the citric acid cycle (Krebs cycle). The oxidation of fatty acids

Figure 6 Different types of starchy food, which are optimal carbohydrate/energy suppliers for intensive sports

(and also amino acids) in the citric acid cycle would be compromised if there was a lack of the required intermediate compounds. (See also fat metabolism in Chapter 3.) In these circumstances the body will start to produce glucose from other substrates (a process called gluconeogenesis) in conditions that lead to a lack in circulating blood glucose (135, 138).

TIME COURSE OF GLYCOGEN DEPLETION

Four important factors determine the speed and the extent to which CHO stores will be emptied.

1. Exercise intensity.
2. Exercise duration.
3. Training status.
4. CHO ingestion.

Exercise Intensity and Duration

As explained above, the use of glycogen depends primarily on exercise intensity and duration. At low to moderate intensities, fat will also serve as a substantial energy source, while CHO reserves will be utilized slowly, for example in a cycling event lasting 4 h, during which exercise intensity approximates 55%–60% of VO_2 max. Also, the relative contribution of fat production will be less during shorter events with a higher intensity, such as a run of 1.5 h at 65% VO_2 max. A maximal contribution of CHO and

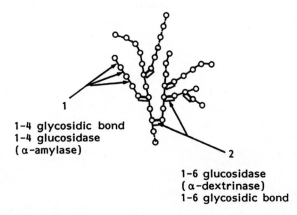

1
1–4 glycosidic bond
1–4 glucosidase
(α–amylase)

2

1–6 glucosidase
(α–dextrinase)
1–6 glycosidic bond

Figure 7 Starch is built up from glucose molecules (circles) which are bound by 1–4 and 1–6 bonds. Both bonds are split by specific enzymes during the process of digestion. Glucose, the end product of starch digestion is absorbed by the gut and delivered to the blood. After uptake by the muscle and the liver, glucose can be restored in the form of glycogen which has a structure comparable to that of plant starch

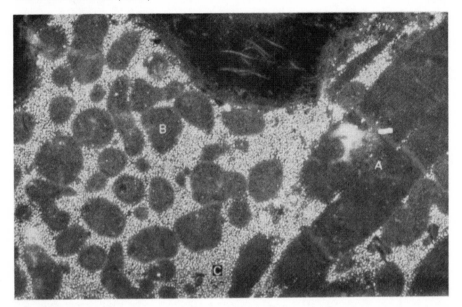

Figure 8 Muscle glycogen is stored as 'starch' granules (C) within the muscle fibres (A) between the mitochondria (B). The amount of glycogen stored can be measured by biochemical analysis of a muscle biopsy sample. Nuclear magnetic resonance, a more modern technique, is non-invasive and makes possible the study of relative changes in glycogen content

relatively low contribution of fat will be present in events that require a maximal exercise capacity, for example, during highly intensive training sessions such as interval and tempo training bouts (44). (VO_2 max = maximal oxygen uptake. Oxygen uptake increases with increasing exercise intensity until a maximum is achieved. The exercise intensity at this point is determined as 100% VO_2.)

Training Status

The time course of glycogen depletion will also be influenced by the training status of the individual. Compared to less well trained individuals, highly trained individuals have an enhanced capacity to mobilize fatty acids from the fat depot, transport these to muscle and use them as an energy source. Thus, when working at the same absolute exercise intensity (e.g. running at a speed of 15 km/h), trained individuals will use less CHO and more fat for muscle contractions (19, 71). Under competition circumstances, however, this may not necessarily be the case as any individual then will work at his or her individual maximal capacity. For

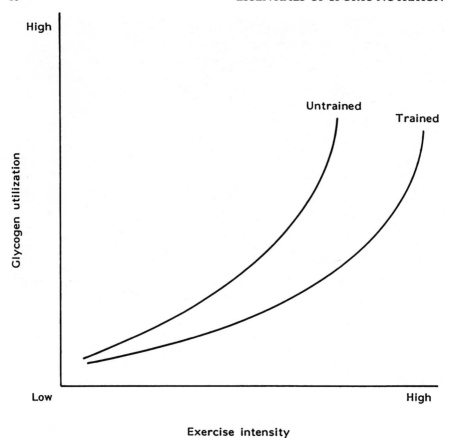

Figure 9 Glycogen depletion rate depends on exercise intensity. Trained subjects use less glycogen and more fat at submaximal exercise intensities. At maximal intensities these differences are less pronounced.

example, the trained runner will run at a speed of 20 km/h while the less trained runner runs at a speed of 15 km/h.

CARBOHYDRATE INGESTION DURING EXERCISE

The rate of utilization of glucose from stored glycogen in the body can be reduced by supplying oral CHO. For example, when food containing CHO is ingested, digested and absorbed, the digested CHO will enter the circulation as the constituent monosaccharides, mainly glucose and fructose. Accordingly, blood glucose rises after oral CHO intake. This rise reduces the need to break down liver glycogen for the maintenance of an

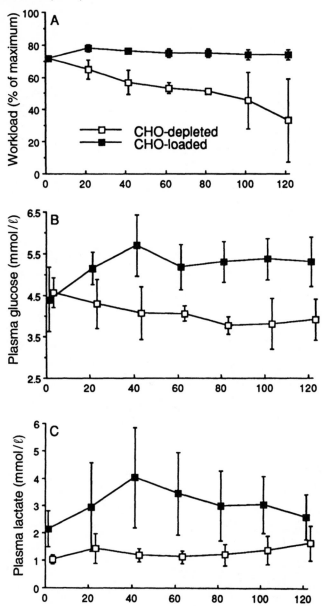

Figure 10 These figures give a typical representation of metabolism in a glycogen depleted state. Because of lack of glycogen in the liver, blood glucose falls, lactic acid production is decreased and fat metabolism is increased to compensate for energy deficits. The consequence is that the performance level will drop to approximately 50% of maximal capacity. Reproduced from Wagenmakers (189) with permission of the American Physiological Society

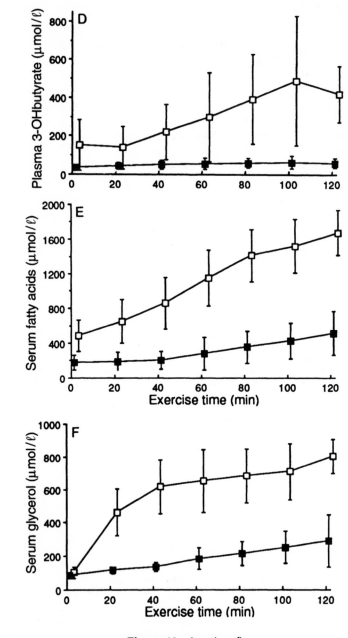

Figure 10   *(continued)*

appropriate blood glucose level. Additionally, glucose supply to and glucose uptake by the muscle will be elevated. Indeed, a large body of scientific evidence shows that oral CHO intake reduces liver glucose output (492) but increases blood glucose at a similar rate. The increased blood glucose after CHO intake will stimulate insulin release and with it glucose uptake by the muscle as well as subsequent CHO oxidation (49, 75, 82, 124, 148).

Theoretically these events will reduce the rate of muscle glycogen and protein degradation for energy production and delay the onset of fatigue/improve performance. Yaspelkis (218) observed that the ingestion of an 8.5% glucose polymer solution reduced the rate of muscle glycogen depletion during low intensity exercise in the heat, while maintaining a high rate of CHO oxidation.

Thus, in studies where CHO was ingested during exercise, total CHO utilization was found not to differ from control groups that did not ingest CHO. Since, in such studies, oral CHO was shown to be oxidized, the conclusion is that glycogen must have been spared. However, since CHO ingestion has not been found to reduce the rate of muscle glycogen degradation in active muscle, this glycogen sparing effect most probably took place in the liver and in non-active muscle (45).

The latter does not necessarily mean that muscle glycogen in active muscle cannot be modified by CHO consumption during exercise. Simply imagine that the supply of CHO during exercise is in excess of the requirement for energy production. In that case the muscle has to store the CHO as glycogen. This would reduce glycogen degradation or even lead to glycogen build-up during exercise. The question is: is that possible? The answer is yes! But, there is one prerequisite to achieving sparing or build-up of endogenous CHO pools during exercise in that the ingested CHO should be easily digested, rapidly absorbed and substantially elevate blood glucose levels.

For exercise lasting longer than 45 min it is recommended that at least 20 g, but optimally up to 60 g, be consumed, with sufficient fluid, during every following hour of exercise (39, 44). Such amounts have been shown not to delay gastric emptying to a physiologically important degree and to stimulate water absorption in the intestine. This aspect is of particular importance in endurance events in the heat, where both CHO and fluid availability may be performance limiting factors (see also Chapter 5). The CHO sources used should be rapidly digestible and absorbable. Most efficient are (soluble) CHO sources which can be ingested with fluid. The gastric emptying rate should be relatively fast and the physical form of the CHO should allow rapid digestion/enzymatic hydrolysis. This is not the case with all CHO sources. For example, the dietary fibre in which some CHO sources are 'packed' may form a physical barrier to digestive enzymes (47) and may also reduce gastric emptying rate. Normal daily meals should primarily contain foods that are rich in slowly digestible CHO and dietary fibre resulting in a low glycaemic index. Examples of such foods are whole grain products and cereals. However, foods taken shortly before and during exercise should be low in dietary fibre and have a high glycaemic index, in order to allow for a rapid gastric emptying and digestion/absorption (30, 44).

The reason for this apparent paradox is that dietary fibre may reduce gastric emptying and decrease the degree in which enzymes can reach the starch for hydrolysis. Fibre also increases gastrointestinal bulk due to water uptake and swelling. Fibre enhances transit in the gut and may be subject to bacterial fermentation causing gas production. Softening of the intestinal contents by fibre and the related improved intestinal transit are desirable in sedentary individuals but may pose a problem during intensive exercise. These factors may explain why athletes who ingest slowly digestible whole grain foods, prior to and during exercise, experience more gastrointestinal problems than athletes who ingest low fibre products (30, 156).

When dietary fibre is excluded from the CHO source, the starch will be fully accessible to enzymatic digestion. The starch and glucose polymers have been shown to be as effective in energy supply as free glucose (75). Other sources of complex CHO, such as rice, spaghetti and potato, are of particular interest for daily CHO intake, between sport sessions, but are shown to be oxidized more slowly during exercise than soluble CHO sources (76). During periods of non-intensive exercise, however, such as mountain walking, these CHO sources can be consumed satisfactorily prior to and also during the activity.

Optimal CHO sources for high intensity endurance events are processed (pre-digested) CHOs that are low in dietary fibre:

- Monosaccharides (glucose)
- Disaccharides (sucrose, maltose)
- Glucose polymers (maltodextrins)
- Starches (suspensible starch).

These types of CHO have the additional benefit of being easily dissolved in fluids, which is an important aspect as the requirements for CHO and fluid (see Chapter 5) are determined by the exercise intensity and duration. The types of CHO listed above have been shown to be about equally effective in increasing blood glucose levels and oxidation rates during exercise as well as in improving performance (43, 44, 82). Effects on blood insulin levels during exercise also do not appear to be different (44).

Some early studies showed that an intake of 50–75 g of rapidly absorbable CHO prior to exercise induces a rapid rise in blood glucose and insulin and a rebound hypoglycaemia as well as decreased performance during the subsequent exercise. However, these studies were done after an overnight fast and CHO was ingested in the resting state,

Figure 11 The difference between 'raw' carbohydrate sources and refined sources is the dietary fibre content. Dietary fibre reduces gastric emptying rate, slows down digestion and absorption and enhances the amount of intestinal bulk, which promotes normal transit

Figure 12 Exercise test under controlled laboratory circumstances. From the breath samples taken it is possible to calculate the carbohydrate oxidation by measuring the content of carbon-13, a naturally stable isotope, in the carbon dioxide

45–60 min prior to exercise. These conditions are not comparable to those of the endurance athlete, who will eat a pre-game breakfast and perform a warming-up prior to the start of the competition. Studies done under real competition conditions did not show any rebound hypoglycaemia (26, 27). Also, CHO intake during exercise will counteraffect pre-exercises diet effects (487).

Meanwhile a large number of studies have shown that pre-game CHO intake can be beneficial in delaying fatigue (for review see 44, 45). Chryssanthopoulos *et al.* (217) compared the effect of ingesting a CHO solution during a 30 km treadmill trial with ingestion of a high CHO meal 4 h prior to the run. Performance times were identical and there were no differences in self-selected running speeds. In both cases, blood glucose was maintained above 4.5 mmol/l, while blood glucose was ±50% higher in the drink trial than in the meal trial. One exception may be that of ingesting pure fructose, which has been shown to maintain a normal blood glucose level while not influencing insulin secretion. This will result in a less potent inhibition of free fatty acid mobilization than after the ingestion of glucose, which will raise insulin levels to an extent that the mobilization of free fatty acids will be inhibited. On the other hand, however, it is known that fructose is passively and thus slowly absorbed. This may lead to intestinal side effects due to fructose accumulation in the intestine whenever the fructose supply exceeds the rate at which it is absorbed. Some authors have reported intestinal upset when >30 g/l have been ingested, both at rest and during exercise (128). However, in one study this effect, with ingestion of up to 1 g/kg body weight during exercise (49), was not found.

Additionally, the rate of fructose oxidation during exercise has been shown to be lower. This may be caused by the concerted action of a relatively slow absorption and by the higher affinity of the enzyme hexokinase for glucose than for fructose. This makes pure fructose as an energy source during exercise less attractive (24, 44, 45, 49, 112, 128). In low concentrations, <35 g/l, or in combination with glucose supplying CHO sources (e.g. glucose, sucrose, maltose, maltodextrins, starch), fructose may not induce gastrointestinal side effects. It has been shown that when fructose is taken in equal amounts with glucose or is taken as sucrose, its absorption is enhanced and is dose dependent (160). The exact mechanism is not known. Also, the addition of an equal amount of CHO to fructose will reduce the delivery from the stomach to the intestine (grams/minute) by about half, thereby reducing the risk of surpassing the slow absorptive capacity. Galactose has been shown to be slowly oxidized and is thus an inappropriate CHO source for supplementation (219).

With respect to palatability and gastrointestinal comfort, starch hydrolysates (maltodextrins/glucose polymers) and dispersible (but not soluble) starch may have the benefit of being less sweet than the mono- and

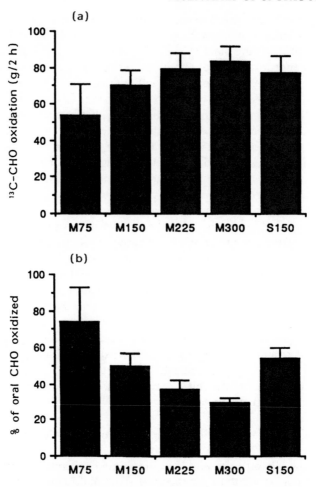

Figure 13 (a) Oxidation of CHO ingested during exercise. (b) Oxidation of oral CHO as a percentage of the given dose (grams). M, maltodextrins (dp20); S, sucrose. Oral CHO is oxidized during exercise and thus contributes to energy production. Increasing CHO intake to about 100 g/h increases its oxidation. Higher CHO intakes have no effect, most probably because of a delayed gastric emptying. This is clearly represented by the lower percentage of CHO oxidized at higher intakes. Interestingly, glucose polymers (maltodextrins) are as well oxidized as sucrose. Reproduced from Wagenmakers *et al.* (213)

disaccharides. They also have less effect on fluid osmolality and have been shown to maximize quantitative glucose absorption, which is of advantage at higher CHO concentrations (155). At higher concentrations (100–200 g/l) drinks would be strongly hypertonic with dissolved mono-/disaccharides but not with maltodextrins/polymers or starch.

CARBOHYDRATE INTAKE AT REST

After exercise the endogenous CHO pools should be replenished. Depending on the time available for total recovery, i.e. the time elapsing between finishing exercise and the next sport activity, there may or may not be a need for speeding up recovery. Glycogen synthesis has been shown to be most rapid during the first few hours after exercise. Thereafter, the synthesis rate will gradually decline (44, 45, 90). Glycogen synthesis itself is only possible if the required building substances, i.e. glucose molecules, are supplied. Net glycogen synthesis rate, therefore, depends on the rate of upregulating synthesis and the quantitative glucose supply (39, 44, 45). The latter depends largely on the type of food ingested, i.e. the rate of digestion and absorption. The CHO source itself may also be important. Glucose favours muscle glycogen recovery, whereas fructose is primarily taken up by the liver, thus favouring liver glycogen recovery (17, 91). When the next activity takes place after one or two days, the athlete can recover properly by ingesting normal meals with a high CHO content, i.e. 55–65 en% (en% = as percentage of total daily energy intake). These meals can best be composed of low glycaemic index foods such as whole grains, cereals, pulses, fruits, vegetables, etc. A relatively slow digestion and absorption rate is favourable in this condition. Under these conditions, 400–600 g of CHO per day should be sufficient to recover glycogen stores for meeting daily energy requirements of up to 4000 kcal (39).

However, if daily energy expenditure is very high, such as during multiday cycling competitions, and exercise intensity is high, the CHO need may reach >12 g/kg body weight per day. In this condition, CHO intake by normal meals composed of low glycaemic index CHO sources may result in too much gastrointestinal bulk and may cause gastrointestinal distress. Therefore, athletes who ingest only normal meals in such circumstances will not be able to ingest enough food, which will result in a negative energy balance and an insufficient CHO intake to compensate for the glycogen used during the exercise. This will lead to starting the next exercise day in a state of incomplete recovery. Sports practice and also controlled experiments have shown that the high needs for energy and CHO during days with an energy expenditure exceeding 4500 kcal/day can only be covered appropriately by the ingestion of CHO foods/solutions with a high glycaemic index (25, 101, 165). (Foods that lead to a slow increase in blood glucose have a low glycaemic index, while those that induce a rapid rise in blood glucose have a high glycaemic index.) Additionally, when the time for recovery is very limited, for example because a second training session or competition will take place on the same day, the food intake in between the normal meals should be composed of foods that are rapidly digested and absorbed, i.e. have a high glycaemic index. Processed, cooked and mashed potato, rice, noodles or corn starch belong to this category. CHO

Figure 14 A six day Tour de France simulation experiment in a respiration chamber at the Maastricht University, Maastricht, The Netherlands. Indirect calorimetry allows continuous measurement of energy expenditure

solutions can be taken during exercise in any situation in which CHO intake through the consumption of normal food cannot take place or is insufficient. This will help to enhance glycogen recovery in the first few hours after exercise (44, 45, 53, 98). An extensive recent review can be found in ref. 489.

PRACTICAL MEASURES FOR GLYCOGEN MODULATION

Based on the known mechanisms and variables that determine the rate of glycogen synthesis and degradation, the following measures can be taken to economize glycogen utilization and maximize exercise performance capacity:

1. Perform regular early morning endurance training at about 50–60% of VO_2 max (heart rate 140–150 beats per minute) on an empty stomach. This will maximize adaptations in fat metabolism, to spare CHO.
2. Enhance glycogen prior to competition by ingesting a high CHO diet followed by a fat rich dinner during the evening prior to competition. This may result in a favourable hormonal milieu and enzymatic activity for reducing CHO oxidation and spare CHO during exercise.
3. Ingest a light, easily digestible mixed pre-game meal containing 40–50% CHO and 30–40% fat. With adequate CHO loading prior to exercise there is no need for a breakfast high in CHO.
4. Do not ingest CHO containing drinks during the last 2 h preceding the competition. Take tea, beverages containing caffeine or plain water in order to maintain low insulin levels to enhance plasma FFA.
5. Perform an appropriate warm-up, in order to raise plasma FFA prior to the start to reduce glycolysis in the early phase of exercise.
6. During exercise one should ingest about 0.5–0.8 g CHO/min along with plenty of fluid (non-hypertonic solution) during the first 90 min of exercise and 0.8–1.2 g/min thereafter, to present a maximal supply of glucose to muscles and liver. During high intensity exercise this may lead to a sparing of liver and non-active muscle glycogen stores, whereas during low intensity exercise periods this may lead to sparing or resynthesis of glycogen in all tissues.
7. Immediately post-exercise CHO should be taken as liquid supplement or in a light, digestible solid form in case recovery between competitions is short.
8. When travelling to a hot or cold climate, significantly different from usual, the athlete should consider appropriate acclimatization in order to reduce catecholamine responses and related effects on glycogen degradation.

(See also Chapter 12.)

Key points

- Carbohydrate (CHO) is the most important nutrient for high intensity performance.
- Energy release from CHO is up to three times as fast as from fat. However, CHO stores in the body are small, which limits the time to perform high intensity exercise.
- Apart from decreasing performance, CHO depletion also induces increased utilization of amino acids (protein) for energy production. This results in the production of ammonia, which may enhance fatigue.
- CHO ingestion during exercise allows sparing of the body's CHO stores, reduction of protein utilization/ammonia production, and a delay of fatigue/improvement of performance.
- Appropriate CHO ingestion between training sessions/days or during intense endurance performance is of importance to avoid progressive fatigue development, reduction in performance capacity and possible signs of overtraining.
- CHO sources to be used during high intensity exercise should preferentially be rapidly absorbable, i.e. have a high glycaemic index, and should be combined with sufficient fluid intake.
- Nutritional interventions shortly before, during and after exercise may negatively affect CHO and glycogen metabolism. Basic knowledge about the timing and the type of CHO intake will be helpful to avoid this.
- CHO supply to optimize performance is not only recommended for long distance runners and cyclists but has also been shown to improve performance in tennis, football and other multiple sprint sports (488).

3   Fat

During physical exercise, skeletal muscle can rely on both fat and carbohydrate (CHO) oxidation to fulfil the need for chemical energy. In resting conditions fatty acid (FA) oxidation contributes considerably to total energy provision. During physical exercise a number of nervous, metabolic and hormonal stimuli will lead to an increased rate of fat mobilization and FA will be increasingly oxidized within the mitochondria of the muscle cells. As a result, the concentration of free fatty acids (FFAs) within the muscle cell will fall, which will stimulate uptake of FFAs from the blood. Increased blood flow to the muscle is the first step in the delivery of more FFAs to the muscle cells. This integrated process of FA mobilization, transport, uptake and finally oxidation is regulated by a concerted action of the hormones adrenaline and noradrenaline (epinephrine and norepinephrine), which increases while exercising, as well as a reduction of circulating insulin. In this respect, the steps to realize an enhanced fat oxidation are numerous and complex. This is the major reason why achieving a steady-state condition of enhanced fat oxidation and related reduced CHO oxidation will take about 20 min. Because energy production from CHO is 'faster' than from fat (119) CHO utilization has to compensate for any shortage of energy that may occur in this initial adaptation phase of slow up-regulation of fat metabolism (1, 136).

Once fat mobilization, transport and uptake are increased, resulting in a metabolic steady state, FFAs from adipose tissue will be available for a very long period. If fat were the only substrate, this would theoretically enable individuals to run continuously at marathon speed for > 70 h, equivalent to an energy expenditure of >70 000 kcal (136).

However, this would only be possible if fat could deliver an adequate amount of energy and if pain in the muscles and joints was not a limiting factor. At maximal endurance competition speeds, CHO availability will be one of the factors limiting performance time, because fat as dominant exercise fuel would be inappropriate in resynthesizing ATP at a high rate (136, 137). With increasing exercise intensity, however, there is a shift to a more pronounced CHO utilization. The idea that the body may use exclusively fat as energy source is thus incorrect. CHO is the prime energy source for the central nervous system and for the red blood cells. CHO availability is also required to ensure that fatty acids can be oxidized in the citric acid cycle. CHO provides the necessary intermediates to keep the citric acid cycle running. For this reason the body will start to produce

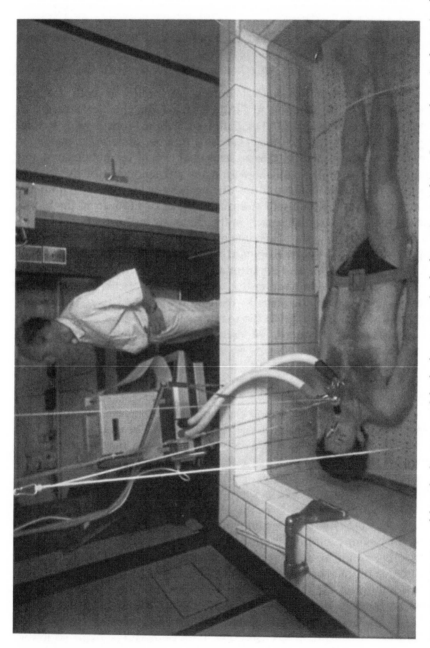

Figure 15 Determination of lean body mass and body fat content by hydrostatic weighing. Lung volume is determined and corrected for by a helium dilution technique (photo R.u.L.)

glucose from other substrates (a process called gluconeogenesis) in any condition that may lead to a lack of blood glucose (135, 138). The relatively low amount of CHO stored in the body poses a limitation for the ability to maintain a high power output during prolonged endurance exercise. Therefore, athletes seek measures that will induce a greater utilization of fat as fuel during exercise, in favour of reducing CHO utilization and, hence, improving endurance capacity. The following paragraphs will describe the mechanisms and regulatory factors involved in the utilization of fat as energy source during physical activity as well as the adaptations that occur as a result of training and dietary intervention.

FAT RESERVES

Fat as energy source has advantages over CHO in that the energy density is higher (37.5 kJ/g vs. 16.9 kJ/g) causing the relative weight of an amount of energy at storage to be lower. CHO stored as glycogen binds approximately 2 g water per gram of glycogen stored (332). This means that changes in muscle glycogen content cause substantial volume effects. As a result, the storage capacity of glycogen in muscle and liver is limited and amounts to approximately 450 g of glycogen in a healthy, untrained male, whereas the storage capacity for fat seems to be almost unlimited.

In non-trained healthy subjects the body fat content may range from 20 to 35% in females and 10 to 20% in males. Fat is stored in the body as triglycerides in fat cells (adipocytes) which make up the adipose tissue. Additionally, a small fraction of triglycerides is stored within muscle cells and a minor fraction of fat circulates in the blood in the form of chylomicrons derived from recently ingested foods and fatty acids bound to a plasma protein called albumin. The major part of adipose tissue can be found under the skin; it is called subcutaneous fat tissue. In addition, fat is stored around the abdominal organs. In highly trained athletes the total amount of fat that is stored in adipose tissue may range from 10 to 25% in females and 5 to 15% in males. This is considerably less than that in sedentary subjects who have a fat content that ranges from 20 to 35% in females to 10 to 20% in males (203). Nevertheless, the relatively low amount of fat stored in the elite athlete has a very large energy potential (approximately 7000 kcal/kg of stored fat). Therefore, adipose tissue serves as the most important energy store that will deliver fatty acids for energy production in all conditions in which, due to a prolonged and insufficient energy intake, the carbohydrate availability becomes limited. This may be the case not only during chronic food deprivation, but also during shorter periods of high energy expenditure resulting in high rate of carbohydrate oxidation and a negative energy (CHO) balance (19, 134, 136, 138).

INTRAMUSCULAR FAT

An alternative source of FAs is triglycerides (TGs) present inside the skeletal muscle cells. Storage as TG takes place in small fat droplets, mainly located in the proximity of the mitochondrial system. Trained muscle may contain approximately 400 g of fat (25, 443, 468).

Release of FAs from muscle-TG is achieved by the action of the enzyme muscle lipase, which is under hormonal as well as local muscular control. Norepinephrine enhances the breakdown of muscle TG whereas insulin counteracts this effect. Apart from hormonal stimuli there is also a local muscular control, shown by the observation that electrical stimulation of muscle enhances TG hydrolysis. Slow twitch muscle fibres have a higher TG content than fast twitch fibres. Endurance exercise has been shown to deplete muscle TG significantly (19, 25, 220). Interestingly, the content of TG stored within the myocyte is increased by regular endurance training.

FAT AS FUEL FOR MUSCLE

Both fatty acids stored in adipose tissue and fat entering the circulation after a meal can serve as potential energy sources for the muscle cell. Moreover, small but physiologically important amounts of FA are stored as triglycerides inside the muscle cells.

The increased activity of the central nervous system will also intensify lipolysis (19, 134, 138). Fatty acids liberated from TGs stored in adipocytes are released to blood, where they are bound to albumin. The albumin transport capacity is in excess of the FA actually bound under physiological circumstances and as such will not be a limiting factor for supply of FA to muscle and the subsequent fatty acid oxidation by muscle.

FA can also be derived from the TG core of circulating chylomicrons and very low density lipoproteins (VLDL) which are both derived from absorbed dietary fat. Chylomicrons are formed in the cell wall of the intestine and enter the bloodstream after passage through the lymphatic system. VLDLs are synthesized in the liver after which they are released directly into the bloodstream.

During the blood flow through the muscle capillaries, fatty acids have to be released from the albumin, the VLDL and the chylomicrons, prior to uptake into muscle. In the case of VLDL and chylomicrons this is achieved by the action of the enzyme lipoprotein lipase (LPL). LPL activity is upregulated by catecholamines and adrenocorticotrophic hormone (ACTH), and downregulated by insulin (220).

After TG hydrolysis, most of the FAs that are liberated will be taken up by muscle, whereas glycerol will be taken away in the bloodstream to the liver where it may serve as a gluconeogenic substrate.

During the post-absorptive state the concentration of circulating TG in plasma is usually higher than that of FA, in contrast to the fasting state when chylomicrons are practically absent in the circulation. Nevertheless, the quantitative contribution of circulating TG to FA oxidation by the exercising muscle cells in humans is uncertain.

FATTY ACID UPTAKE BY MUSCLE

It is generally accepted that the arterial FA concentration strongly affects FA uptake into muscle at rest and during low intensity exercise (220).

During transport of FA from blood to muscle several barriers have to be passed. Each of these barriers may theoretically limit FA uptake and subsequent oxidation by muscle. The following barriers have to be considered: (i) the membranes of the vascular wall (endothelium); (ii) the interstitial space between endothelium and muscle cell; (iii) the membrane of the muscle cell; (iv) cytoplasm of the muscle cell; and (v) mitochondrial membrane.

Fatty acid binding and transporting proteins play a key role in the transport of FA from blood to the mitochondria, where finally L-carnitine has a key function for the transport of FA into the mitochondrion. It is assumed that the latter is not rate limiting for FA oxidation in muscle. Instead, based on the available evidence, it is suggested that the uptake of FA from blood into muscle is the most limiting factor in overall FA utilization during exercise (220).

FATTY ACID OXIDATION BY MUSCLE AND POSSIBLE LIMITATIONS

In the resting muscle cell a relatively high percentage of the overall energy production stems from FA oxidation. This high contribution is either maintained or becomes slightly reduced during light aerobic exercise. However, with high exercise intensities there will be a more pronounced shift from fat as energy source to CHO, particularly at intensities above 70–80% of VO_2max. This points to the fact that there are limitations to increase the FA oxidation rate in order to replenish sufficient ATP to cover the needs. Several theoretical explanations have been given for this exercise induced shift from fat to CHO:

1. An increased rate of glycogen degradation and glycolysis, as observed during high intensity endurance exercise, also enhances lactate formation. Lactate has been observed to reduce lipolysis. The net result will be a decrease in plasma FA concentration and, hence, supply of FA to muscle cells. As a consequence, enhanced carbohydrate oxidation will most likely compensate for the reduced FA oxidation.

2. A lower adenosine triphosphate (ATP) production rate per unit of time from fat compared to CHO, as well as the fact that more oxygen is needed for the production of a certain amount of ATP from fat compared to CHO (119).
3. Limitations in the FA transport from blood to the mitochondria. As mentioned earlier, this transport is influenced by a number of barriers that have to be passed, of which the transport from the blood into the muscle seems to be most limited.

Mitochondrial FA oxidation rate depends on the actual capacity of the carnitine transport system.

It follows from the above paragraphs that the oxidation rate of FA is mainly the mutual result of three processes: (i) lipolysis of TG in adipose tissue and circulating TG and transport of FA from blood plasma to the sarcoplasm; (ii) availability and rate of hydrolysis of intramuscular TG; and (iii) activation of the FA and transport capacity across the mitochondrial membrane.

Furthermore, the processes outlined under (i) and (ii) primarily pose the limitations to fat oxidation observed during maximal FA flux. This is most evident during both short-term intense exercise or during the initial phase of a long-term exercise. In this condition lipolysis in adipose tissue and in

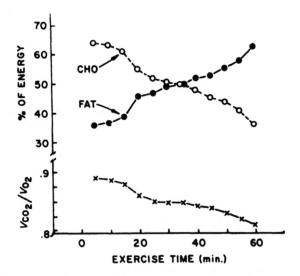

Figure 16 Fat mobilization, transport and utilization are relatively slow processes. At the onset of physical exercise most energy comes from CHO metabolism. After approximately 20 min fat metabolism proceeds at full speed and CHO utilization will be reduced. From D. Costill, *J. Appl. Physiol.* 1979, 47: 787–791. Reproduced with permission from Association Colloques Physiologie, France

muscle-TG is insufficiently upregulated to result in enhanced FA supply. The result will be that the rate of FA oxidation exceeds the rate at which FA is mobilized, leading to a fall in plasma FA and intracellular FA in muscle. As a consequence, the use of CHO from glycogen must be increased to cover the increased energy demand (469, 470).

The extent to which limitation in FA transport and oxidation must be compensated by an enhanced capacity to utilize CHO also becomes transparent when the capacity to oxidize FA is analysed in different muscle fibres. There is a clear functional relationship between fibre type, microstructure, substrate stores and CHO or FA oxidation capacity. Slow twitch muscle fibres have a relatively high degree of capillarization, a high fatty acid binding protein (FABP) content, a high mitochondrial density and a high muscle lipase and intracellular TG content, which are associated with a high FA oxidation capacity. Fast twitch muscle fibres, on the contrary, are low in all these factors, i.e. they are extremely limited in the ability to oxidize FA. Therefore, these fibres must rely primarily on CHO as exercise fuel.

STRATEGIES TO IMPROVE FATTY ACID OXIDATION

Several recent reviews have described in detail the effects of exercise on fat metabolism as well as the effects of various methods to modify fat metabolism in the athlete (220–222). The most important aspects are outlined below.

As pointed out earlier there is a progressive shift to the use of CHO oxidation with increasing exercise intensity. This has its origin in stronger metabolic and hormonal responses which induce an enhanced glycogen breakdown and lactate formation, as well as progressively increased recruitment of fast twitch muscle fibres, which generally lack the capacity to oxidize substantial amounts of FA.

Since the storage of CHO in the form of glycogen is limited, the ability to perform high intensity exercise will be decreased with progressive glycogen depletion (332). Any adaptation leading to an increased capacity to use FA for ATP resynthesis will lead to a sparing of endogenous CHO with the consequence that endurance capacity may be improved. Theoretically there may be a number of intervention possibilities to increase plasma FA levels and to improve the mechanisms involved in transport and oxidation of FA. Most of these interventions have been studied over the last three decades:

- Training
- Medium chain triglyceride feedings
- Oral fat emulsions and fat infusions
- Caffeine

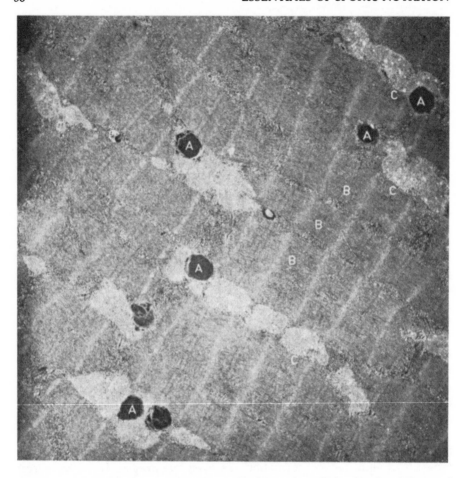

Figure 17 Muscle tissue contains fat in the form of small fat droplets (A) stored within muscle fibres (B), near mitochondrial networks (C). Endurance trained athletes possess more intramuscular fat

- L-Carnitine supply
- High fat diet.

PHYSICAL TRAINING

Endurance training has been observed to result in a number of structural and metabolic adaptations, which will favour FA oxidation. Whereas α-adrenergic mechanisms regulate lipolysis at rest, β-adrenergic activity has been found to determine lipolysis during exercise (430). The sensitivity of β-

adrenoceptors for catecholamines in the adipocyte will increase as a result of exercise (467). Sensitivity may be further enhanced as a result of adaptation to regular training. In addition, training will help fat cells to increase their sensitivity to stimuli for FFA mobilization, thereby improving the speed of adaptation to the higher needs when exercising (19). During maximal exercise intensity, however, the hormonal and metabolic stimuli to enhance CHO mobilization and along with it the mobilization, uptake and utilization of FAs, are maximized. In this condition the resulting increase in blood FFAs does not automatically lead to a reduction of muscle and liver glycogen utilization (7, 162). This will theoretically promote the delivery of FA from the fat cells to the blood. However, recently it was shown by Romijn *et al.* (461) that the rate of appearance of FA from adipose tissue is decreased in the trained individual.

The capillary density of muscle tissue will increase, which in itself augments the exchange surface area, promotes blood flow and with it the delivery of oxygen and FA (437, 438). Training also induces an increase in sarcolemmal fatty acid binding protein, which contributes to the translocation of FA into muscle (452). Within the muscle cell there will be an increased mitochondrial volume as well as mitochondrial enzyme activity (463).

Trained muscles express a higher activity of LPL, muscle lipase, fatty acyl CoA synthetase and dehydrogenase, carnitine-acyl transferase and 3-hydroxyacyl CoA dehydrogenase, which will be in favour of enhancing FA supply to the mitochondria and subsequent oxidation (463). As a result, trained muscles are able to oxidize more substrate (438), which is also expressed in increased oxygen consumption at maximal exercise intensities (436, 457).

Lastly, trained muscles store more intracellular fat in lipid droplets located along the surface of the mitochondrial system; they may theoretically enhance the capacity to supply and oxidize FA derived from the intracellular lipid store (463).

Increased intracellular TG storage as well as observations from arteriovenous difference and isotope labelling experiments indicate that highly trained endurance athletes rely more on the utilization of intramuscular stored FA during exercise and less on the utilization of blood-borne FA (220, 463). The advantage of a shift from extracellular to intracellular stores of FA is that some potential barriers in overall FA utilization, such as the endothelium and the sarcolemma, are irrelevant when intracellular TG is utilized.

Thus, training enhances total FA oxidation, especially by increasing intramuscular fat storage and maximal FA flux. Along with this, endogenous CHO stores will be conserved during exercise in the endurance trained individual, which prolongs the time period during which intense exercise can be performed.

MEDIUM CHAIN TRIACYLGLYCEROL (MCT) INGESTION

MCT contains fatty acids with a chain length of six, eight or ten carbon atoms. Generally, MCT is rapidly emptied from the stomach and taken up by the intestine (471). After absorption by the enterocyte MCT is transported with blood to the liver, in contrast to long chain triacylglycerol (LCT) which is transported by the lymphatic system to the vena cava. MCT readily increases plasma medium chain FA and TG levels. In muscle, medium chain FA is rapidly taken up by the mitochondria, not requiring the carnitine transport system (22). Consequently, MCT is oxidized readily (7, 48, 111) and faster than LCT (435). This has led to the assumption that MCT may be an effective exogenous fuel for exercising muscle and that MCT ingestion may potentially enhance fat oxidation and thereby reduce CHO utilization.

Early studies have indicated that oral MCT, taken shortly before exercise, is only partly oxidized during exercise and has not been shown to improve performance (7, 92 162). In a study by Ivy et al. (92) 30–60 g of MCT were ingested with a cereal meal 1 h prior to exercise. Most probably because of the relatively low oxidation of the oral MCT, no differences in CHO oxidation were found. In two other studies there was a substantial oxidation of the ingested MCT (48, 111, 455). However, in these studies the amounts of MCT ingested were relatively small. Unfortunately, the effect of MCT feedings on performance was not measured in any of these studies.

More recently several stable isotope studies have been performed to evaluate the effect of MCT or MCT + CHO ingestion on exogenous, endogenous and total fat and CHO oxidation. These studies have shown that oral MCT is rapidly oxidized by muscle but does not lead to glycogen sparing in active muscle cells as measured from muscle biopsy specimen (446–449). The fact that total fat oxidation remained the same after MCT ingestion, even in a glycogen depleted state (446), points to the fact that oral MCT most likely competes with long chain FA and, hence, leads to a sparing of endogenous fat stores, probably intramuscular fat. This may also explain why no endogenous CHO sparing took place. Also in the studies of Jeukendrup et al. (446–449) relatively small amounts of MCT were supplied to the athletes. The background of this small supply was that ingestion of >30 g in a short period of time induces nausea and gastrointestinal discomfort. It may be speculated that this may be caused by a relatively high cholecystokinine (CCK) release after MCT intake (432).

In a recent study by van Zyl et al. (464), however, subjects ingested 86 g of MCT during submaximal endurance exercise lasting 2 h, followed by a 40 km time trial. Ingestion took place as 4.3% w/v MCT drink, 10% w/v CHO + 4.3% w/v MCT drink or 10% w/v CHO drink as control. Interestingly, they observed the poorest performance with ingestion of MCT alone but a significantly improved performance with the CHO + MCT trial compared to the CHO trial. No mention was made of any gastrointestinal discomfort. The authors did not measure muscle glycogen but speculated on the basis of a reduced endogenous CHO oxidation that glycogen may have been spared and that this might explain the performance benefit observed. These findings are in contrast to the earlier mentioned observations by Jeukendrup et al. (446) who observed no endogenous CHO or glycogen sparing. This has prompted Jeukendrup et al. (449) to perform a similar experiment in which the subjects ingested 85 g MCT as MCT drink, CHO + MCT drink or for control a placebo drink during an endurance exercise lasting 2 h at an intensity of 60% VO_{2max}, followed by a 15 min time trial. In this particular study the performance test was not interfered with by any physiological measurement. In contrast to the study of van Zyl et al. (464), performance was not improved by the MCT + CHO treatments. A substantial number of subjects experienced gastrointestinal problems with MCT ingestion. The reason for the discrepancy in the data of these studies remains unclear.

Thus, from the available data it cannot be concluded that MCT ingestion is of benefit for glycogen sparing and/or improving endurance performance.

ORAL FAT AND FAT INFUSIONS

Another attempt to improve fat oxidation has been to enhance the blood long chain FA levels by infusing lipid emulsions. This procedure has been shown to result in a significant reduction of glycogen degradation in two studies (445, 466). In line with the positive effects of fat infusion on muscle glycogen sparing, the opposite—a decline of plasma FA, induced by inhibiting lipolysis by nicotinic acid—resulted in an increased rate of muscle glycogen degradation (431). An elevated level of circulating FA is thus a prerequisite for reducing the rate of endogenous CHO utilization during exercise. However, for sports practice this procedure seems to be impractical. Infusions during competition are not possible and even if they were, the IOC doping regulations, which consider any artificial measure to enhance performance as unethical, would forbid them.

Oral intake of fat emulsions may not be of benefit either. Oral fat may inhibit the gastric emptying rate of rehydration solutions also ingested during exercise and may lead to gastrointestinal discomfort (30). Additionally, it will take a considerable time before the absorbed long chain triacylglycerols will be available for oxidation because of passage through the lymphatic system.

To our knowledge there are currently no studies that have convincingly shown any benefit of fat ingestion shortly before or during exercise. One study (490) pointed to a favourable effect of ingesting a high fat meal 3 hours before exercise, in combination with a heparin infusion (improving fatty acid mobilisation). The latter, however, would fall under the doping regulations. A study with a high fat meal alone (406) showed no effect. Thus, although oral supply of fat may increase the blood fat level, the uptake of fat from blood into active muscle cells may not be enhanced due to limitations in the FA transport capacity (19, 220, 463). In this respect, fat supply during exercise in trained subjects during competition conditions has not been shown to be of benefit for a reduction of muscle and liver glycogen utilization (7, 162).

CAFFEINE

Caffeine is known to affect muscle, adipose and central nervous tissue indirectly by mediating the level of cyclic adenosine monophosphate (cAMP) and its related calcium release from the intracellular storage sites (220). This effect is initiated by binding of catecholamines to beta-receptors of cell membranes, thereby enhancing the activity of the enzyme adenylate cyclase, which catalyses the formation of cAMP from ATP. Caffeine has been observed to enhance plasma norepinephrine and epinephrine levels. Additionally, caffeine inhibits phosphodiesterase which degrades cAMP to the non-active compound 3'5'-AMP. In this way caffeine increases cAMP half-life and with it lipolysis. By these actions caffeine increases the cAMP level which maximizes the activity of the intra-adipocyte lipase and, hence, lipolysis.

Nevertheless, caffeine has been observed to enhance plasma FA in many studies in man and animals (220, 327, 332). In contrast, an increased fat oxidation (by assessment of the respiratory exchange ratio, RER) and reduced glycogen degradation were observed in only a few of these studies. This may indicate that the caffeine-induced elevation of FA simply comes on top of the relatively high exercise-induced increase in FA, which most likely already maximizes FA transport across the epithelium. These data also indicate that the performance-enhancing effects of caffeine (see also pages 149–159) are most probably related to effects on the central nervous system rather than to effects on fat oxidation and glycogen sparing.

There are reasons to hypothesize that caffeine ingestion may indirectly also counteract its effect on lipolysis and subsequent FA oxidation during exercise. Increased liver glycogen breakdown and plasma lactate levels have been observed after caffeine ingestion (220) and lactate is known to be a strong inhibitor of lipolysis (439). Thus it cannot be excluded that caffeine might also exert depressing effects on FA oxidation in exercising muscle cells.

L-CARNITINE

In humans carnitine is obtained from the diet, especially from red meat. Additionally, carnitine is synthesized in the body from intracellular trimethyllysine, which requires methionine for the methylation process. This biosynthetic process occurs mainly in liver and to a smaller extent in kidney and brain (442) after which L-carnitine is released into the circulation from which it is taken up by muscle. L-Carnitine is lost daily in small amounts from the body via urine and stools. The primary function of L-carnitine is the transfer of long chain FA across the mitochondrial membrane (434), to enter the oxidation pathway.

Addition of L-carnitine to the incubation medium has been shown to markedly enhance the long chain FA oxidation of isolated mitochondria (434). This has led to the speculative assumption that oral L-carnitine intake should lead to enhanced fat oxidation in athletes or in people wanting to lose weight. However, there is no solid scientific evidence that this is the case, despite the enormous amount of positive performance claims made in advertisements for this nutritional aid, as under normal conditions tissue carnitine levels are relatively high and do not form a constraint on FA oxidation.

Oral L-carnitine has been observed to increase the plasma L-carnitine level while uptake in muscle remained unchanged (462). This observation fits well with the finding that L-carnitine is taken up against a concentration gradient—plasma 40–60 μmol and muscle 3–4 mmol (433). This gradient is so large that even a substantial oral intake would not result in a measurable change in this situation. As a result of increased plasma levels and unchanged muscular uptake, urinary carnitine excretion increases many-fold (190).

Additionally, there are no indications that heavy exercise results in a substantial loss of carnitine from muscle cells. No differences in resting carnitine levels have been observed between training and non-training individuals (472). These data, as well as those of other well-controlled recent studies (473–475), failed to show an effect of L-carnitine supplementation on FA oxidation of muscle during exercise (also see pages 145–146). For complete review see Wagenmakers (190, 264) and Heinonen (478).

HIGH FAT DIET

High fat diets are claimed to enhance the capacity to oxidize FA and have attained considerable interest as a potential tool to improve performance in endurance athletes. In rats a high fat diet has been observed to increase LPL activity significantly, compared to animals fed a high CHO diet (460). However, this observation has to be interpreted with caution and may be explained by a strong upregulation of LPL activity with the used combination of *high fat–low CHO* in one group and a downregulation in the other group, receiving *high CHO–low fat*. Thus, most likely, such a striking difference may not appear when a high fat diet is compared to a normal mixed diet.

An increased LPL activity as well as an increased deposition of intracellular fat in muscle may explain a greater availability of FA to the mitochondria after a high fat diet and also may explain the lower RER (220). In rats a high fat diet also induced an improved performance (456). However, there may be significant species differences in FA handling. As such, human studies are of critical importance in order to draw any conclusions.

Johannessen et al. (450) studied seven male subjects who ingested a high fat diet in either solid or liquid form (76 en% fat) during 4 days, or a high CHO diet (76 en% CHO). This diet regimen was followed by a run endurance test till exhaustion. The running test consisted of alternating blocks of 30 min running followed by 10 min rest. Performance was significantly reduced (by approximately 40%) after this short-term high fat diet. Jansson and Kaijser (444) investigated the effect of a high fat diet lasting 5 days (69 en% fat) followed by a 5 day high CHO diet (75 en% CHO) on muscle substrate utilization in 20 subjects. FA utilization was estimated by measuring arteriovenous differences and by measurement of substrate concentrations in muscle biopsy specimens. Although they observed a lower RER after the fat diet and an increased FA extraction by muscle, there was no consistent effect on muscle glycogen utilization. The study included both males and females, was not randomized in treatment order and the diet duration was very short. No performance measures were taken. Phinney et al. (459) studied five cyclists who had to perform an endurance capacity test till exhaustion after a high fat diet lasting 4 weeks. The authors claimed that high fat diet caused a significant improvement in performance. However, the individual performance data show that only two out of five cyclists improved their performance, one of these two by 57%! Two showed a decreased performance and one cyclist remained on the same level. That the overall result was positive was largely on account of the single subject who showed the rather unrealistic 57% performance increase after one month on a high fat diet. Furthermore, no crossover design was used in this study. Lambert et al. (453) studied five well trained cyclists for a period of 14 days who ingested either a high fat diet (67 en% fat) or a high CHO diet (74 en% CHO). The high fat diet led to a reduction of muscle glycogen content of approximately 50% (121 ± 4 and 68 ± 4 mmol/kg w/w for high CHO and high fat treatment respectively). In a high intensity cycling test to exhaustion (85% VO_2max) there were no statistically significant differences between the treatments, although the mean values were quite different in terms of athletic performance times (8.3 ± 2.3 and 12.5 ± 3.8 min for high fat and high CHO diet respectively). During a low intensity performance trial, which followed the high intensity trial after a rest period of 20 min, time to exhaustion was significantly prolonged. However, despite the fact that exhaustion occurred, the heart rate observed was only 142 ± 7 beats/min in the high fat diet and 143 ± 8

beats/min in the high CHO trial, in contrast to a heart rate of >180 beats/min in the high intensity trial. The fact that the preload (the high intensity test) was not standardized and that heart rate response does not reflect the stress of exercise-induced exhaustion points to the possibility that variables other than the difference in the diet alone, e.g. motivational, may have influenced the performance results. The very large difference in time to exhaustion in the low intensity (50% peak power output) trial (79.7 ± 7.6 min vs. 42.5 ± 6.8 min for high fat and high CHO diet respectively) further underlines this suggestion. It can be questioned whether such a large performance difference can be caused by 14 days of a high fat diet alone.

Muoio *et al.* (458) tested runners on a treadmill after a diet intervention lasting 3 weeks and observed a significant increase in time to exhaustion from 76 to 91 min while running at an intensity of 75–85% VO_2max. Moreover, the 'high fat diet' consisted of 50 en% CHO and 38 en% fat. As such, this is comparable to a normal mixed diet consumed by many athletes. Thus, since there was no real high fat diet and no change in fat oxidation was observed, it is unclear whether this performance capacity improvement is the result of fat in the diet.

Hoppeler *et al* (485) compared the effect of a low fat diet (18.4%) to a high fat diet (40.6%) on endurance capacity during a run at 80% VO_2max until exhaustion. VO_2max remained unchanged but endurance capacity improved by 21%. It remains to be established whether such an effect also improves the performance to run a certain distance faster.

Helge and Kiens (441) studied the effect of combined training and diet on performance progression in 20 untrained subjects, divided into two groups of 10. These subjects performed endurance training for a period of 7 weeks, three or four times per week, while ingesting either a 65 en% CHO or a 62 en% fat diet. This period was followed by another training period of 1 week, while ingesting the CHO rich diet alone. The results showed that maximum oxygen consumption increased by 11% in both diet groups. Performance progression, however, was significantly better with the high CHO diet. From 35.2 ± 4.5 min to 102.4 ± 5.0 min in the high CHO diet and from 35.7 ± 3.8 min to 65.2 ± 7.2 min in the high fat diet, respectively. After the final week on the CHO diet, the performance improvement in the previous CHO-treated groups was maintained while the previous fat diet group further improved endurance performance from 65.2 ± 7.2 min to 76.7 ± 8.7 min. This, however, was still below the achieved performance of the CHO diet group; i.e. 103.6 ± 7.2 min. Heart rate and noradrenaline levels were highest while being on the high fat diet.

These results indicate that a high fat diet is detrimental with respect to training and performance progression at the beginning of an endurance training programme. This was the case despite the observation that the high fat diet resulted in a 25% increase in β-hydroxyacyl-CoA-dehydrogenase, one of the key enzymes in FA oxidation, in the high fat diet group compared to no change in the CHO diet group. It should be emphasized, however, that these data do not allow for a generalization towards highly trained individuals.

To the best of our knowledge no other human studies on the effect of high fat diets are available at this moment. Seen against the bulk of the evidence that CHO ingestion improves endurance performance tasks, it remains speculative to state that a high fat diet, which downregulates CHO metabolism as well as may decrease glycogen stores in muscle and liver, may lead to better results. The fact that high fat diets are unpalatable restricts most attempts to study their effects in humans to a duration of several weeks at maximum. On the one hand this may be too short to achieve measurable adaptation effects. On the other hand, long-lasting trials may result in adverse health effects on the cardiovascular system as well as lead to insulin insensitivity, especially in less well trained subjects, due to overexposure of lipids to the body. Interestingly, in the available human fat diet performance studies, no systematic measurement of changes in lipoproteins were undertaken. Recently, Leddy et al. (454) reported data on 12 male and 13 female runners who, divided in subgroups, raised daily fat intake from 16 to either 30 or 40 en%, for 4 weeks. This increase in fat was not associated with changes in LDL cholesterol, apolipoprotein B or Apo A1/Apo B ratio, but raised HDL cholesterol. This study indicates that shifting from a high CHO diet to a diet that has a fat content comparable to that of many sedentary individuals, is not associated with negative side effects for well trained athletes. Since most high fat diets tested and sometimes recommended to athletes have a substantially higher fat content, i.e. 50–65 en%, additional studies are required to evaluate the possible effects on cardiovascular risk factors.

The recent findings by Van Zyl et al. (465) point to the fact that a combination of a short-term high fat diet followed by a high CHO diet may improve endurance performance. They studied five trained cyclists who ingested in random order either their habitual diet or a high fat diet (65 en%) for 10 days, followed by a 3 day high CHO diet (65 en%). This dietary preparation was followed by an exercise preload of 2.5 h at an intensity of 70% VO_{2max}, after which a 20 km time trial was performed. During the time trial the subjects ingested a CHO–MCT suspension. Time trial performance was improved by 80 s ($p < 0.05$) after the 7-day high fat and 3-day high CHO preparation. However, more studies and with a greater number of subjects need to be done before any well founded recommendations on this type of nutritional preparation can be made. No performance effects were observed by Burke (484) in 8 trained cyclists after a 5 day high-fat diet. Interesting in this regard are the observations made by Helge et al. (440) that a high fat diet lasting 7 weeks, irrespective of training, increases β-hydroxyacyl-CoA-dehydrogenase activity in muscle

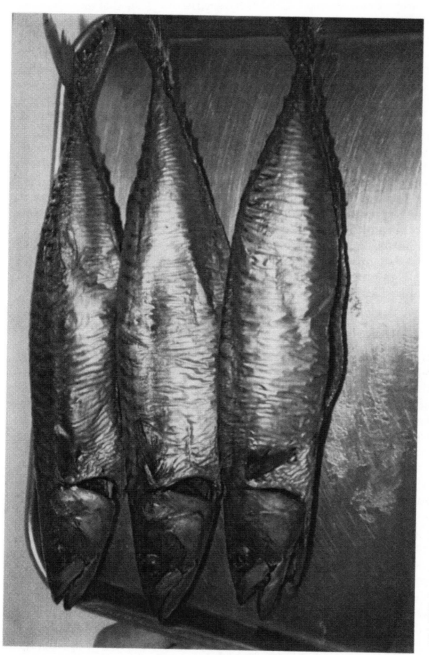

Figure 18 Fish is an excellent source of polyunsaturated fatty acids. Omega-3 fatty acids are important for membrane plasticity and stress tolerance of red blood cells

and that this effect does not occur in subjects following the same training programme but ingesting a CHO rich diet. This suggests that diet *per se* can influence endurance exercise-induced adaptations in muscle. Thus, from the paragraphs above it appears that although a number of intervention possibilities to enhance FA oxidation during exercise—with the goal to improve endurance capacity—have been studied, only regular endurance training can be classified as being successful in this respect. Although some very recent data after following a combined dietary intervention (CHO rich diet→short-term high fat diet→high CHO diet→competition) show improvements in performance during low intensity exercise, the bulk of evidence points to the fact that high intensity exercise performance is best achieved after being on a diet which is relatively high in CHO and low in fat. One other aspect should be mentioned. A substantial number of animal studies has shown that high fat diets result in insulin resistance and type II diabetic responses (43). The safety healthyness of a high fat diet for athletes has not been established in this respect.

Statements that L-carnitine, caffeine, MCT feedings, oral TG feedings and high fat diets may improve endurance performance of endurance athletes during high intensity events can at present not be supported by consistent and solid scientific evidence.

FAT INTAKE

Sedentary people living in industrialized countries consume diets that contain 35–45% of total energy content as fat (19, 131). These figures are relatively high, seen in the light of recommendations that daily food should be rich in CHO (>50 en%). Athletes are generally advised to reduce fat intake to approximately 25–35 en%, thereby enhancing CHO intake to 55–65 en% (19, 39, 44, 45, 90, 173). This reduced fat intake should to a large extent be realized by consumption of lean meat and low fat foods. Saturated fatty acid intake should be limited to less than 10 en%, mainly by making use of plant oils for meal preparation instead of hard saturated fats. With improved food quality and increased total energy consumption, the low figure of 25–35 en% fat intake in athletes will lead to a more than sufficient supply of the essential fatty acids that are required for normal biological functions (at least >1 en%, preferably about 7 en% should be in the form of mono- and polyunsaturated fatty acids (131)). Polyunsaturated fatty acids are known to influence the structure of the cell membrane especially of red blood cells. A group of French scientists (77) reported that an increased intake of omega-3 fatty acids, by means of supplementation (e.g. fish oil capsules), resulted in improved red blood cell plasticity, maximal oxygen consumption and better blood oxygen levels when exercising at high altitude. However, similar findings have never been reported when exercising at sea level. A recent study (223) performed at sea level did not result in any performance benefit. In this study fish oil rich in omega-3 fatty acids was given for a period of 3 weeks at 6 g daily. A significant increase in the amount of polyunsaturated fatty acids in the cell membranes of red blood cells was observed but the deformability of the red blood cells during exercise remained unchanged.

Key points

- Fat is a 'slow' energy source compared to CHO.
- When primarily using fat as an energy source, athletes can only work at 40–60% of their maximal work capacity.
- Increased fat utilization, as a result of training, reduces the use of CHO from the glycogen stores in the body and will accordingly influence the duration of sufficient CHO availability during exercise. The latter will have an impact on muscle fatigue when exercise is intensive.
- The daily fat intake in athletes is recommended to be relatively low, i.e. <30 en%, allowing for an increase in the proportion of CHO in the diet. Saturated fat sources should be avoided and foods rich in or prepared with oils that are rich in mono- and polyunsaturated fatty acids, such as vegetable and fish oil, should be promoted.
- Some very recent data show improvements in performance during low intensity exercise after following a combined dietary intervention: CHO rich diet → short-term high fat diet → high CHO diet → competition. However, the bulk of evidence points to the fact that high intensity exercise performance is best achieved after being on a diet which is relatively high in CHO and low in fat.
- Statements that L-carnitine, caffeine, MCT feedings, oral TG feedings and high fat diets may improve performance of endurance athletes during high intensity events, by means of boosting fat metabolism, can at present not be supported by consistent and solid scientific evidence.

4 Protein

An appropriate protein supply with the daily diet is essential for growth and development of organs and tissues. Muscle hypertrophy requires amino acids; an insufficient supply of protein in general or of essential amino acids (those which cannot be synthesized by the human body) in particular is known to be associated with impaired growth. In the subsequent paragraphs we will briefly describe how key biological functions depend on an appropriate protein supply and how these are influenced by exercise. For a schematic presentation of protein metabolism see Chapter 13.

PROTEIN RESERVES

The human body has no protein reserve comparable to the large energy store in adipose tissue and glycogen. All protein in the body is functional protein, i.e. it is either part of tissue structures or part of metabolic systems such as transport systems, hormones, etc. Any abundant protein cannot be stored as protein. Accordingly the body will degrade the non-used protein, oxidize the liberated amino acids and excrete its nitrogen with urine. Alternatively the amino acids can be metabolically converted into either glucose or fatty acids that can be stored in the respective pools. In conditions of energy deficits, amino acids may be used primarily as energy fuel to resynthesize ATP (25, 63, 64, 71, 106, 107, 125, 138, 139, 157, 187, 188).

The human body possesses three major functional protein pools:

1. The plasma proteins and plasma amino acids (AAs)
2. Muscle protein
3. Visceral (abdominal organs) protein.

PLASMA PROTEINS/AMINO ACIDS

Albumin and red blood cells are important plasma proteins. Both are involved in transport processes (carriers) and may be reduced as a result of long-term insufficient protein (nitrogen) intake, energy intake, or a combination of both. Other plasma proteins with a rapid turnover, such as pre-albumin and retinol binding protein, respond to short-term changes and have therefore been used as markers for nutritional status (161). Since transport of oxygen by haemoglobin plays an essential role 'to feed' the metabolic chains of energy production, it may be concluded that any significant reduction in haemoglobin will be associated with impaired metabolism and performance (38, 209). In contrast, boosting the haemoglobin level by the use of EPO is known to enhance performance capacity and is therefore regarded as unethical and doping.

The circulating plasma AAs make up the central pool of metabolically available protein substances. Any protein consumed will, after digestion and absorption, feed into the plasma AA pool. Any AAs for synthesis of functional protein will be taken up from this AA pool.

The composition of the plasma AA pool is kept within a narrow range. Shortage of non-essential AAs will induce production of these AAs by the body (*de novo* synthesis). Shortage of essential AAs on the other hand cannot be compensated by *de novo* synthesis. There are only two ways to compensate for such a shortage: increased consumption of protein containing these essential AAs or breakdown of functional protein within the body. The latter will lead to the liberation of AAs into the plasma pool. Apart from being the building bricks of all tissues, circulating AAs have also a large number of key functions in energy metabolism and the central nervous system. AAs play a major role in intermediate metabolism; they are precursors for gluconeogenesis and hormones as well as peptides, which function as neurotransmitters (64, 71, 80, 99, 106, 107). Any pronounced change in plasma AA composition can therefore affect protein synthesis rate, alertness, fatigue, mood, etc. (206). Any prolonged change may also have consequences for health.

Influence of Exercise

Exercise is known to be associated with changes in plasma AA composition. It has been shown that branched chain amino acids (BCAAs: leucine, valine, isoleucine) by being oxidized contribute to energy production during exercise. As a result, their concentration in plasma will fall (1, 106, 107, 187, 188). This has two important consequences: (i) the oxidation of BCAAs will lead to the formation of ammonia, a metabolic end-product in principle known to be toxic and to be associated with fatigue (29, 187, 188); and (ii) the ratio of BCAAs to other amino acids will change. Such a change will lead to an increased transport of some AAs, e.g. tryptophan, which are known to be precursors of hormones and peptides in the central nervous system, into the brain. This changed amino acid uptake is thought to influence neurotransmission and fatigue (139).

It has been shown that a shortage of CHO (glycogen, blood glucose) dramatically increases the use of protein (BCAAs) for the production of energy (106, 107, 188). Two lines of evidence support this finding:

1. Depletion of endogenous CHO pools leads to:
   (a) dramatic changes in intramuscular and plasma AAs;

(b) increases in the activity of enzyme complexes involved in the breakdown and oxidation of BCAAs;

(c) rapidly increasing intramuscular and plasma ammonia levels;

(d) a reduction of the time to exhaustion;

(e) increased nitrogen loss through sweat and urine.

2. Supplementation with CHO, maintaining sufficient available endogenous CHO, minimizes these changes (187–189).

Exhausting athletic effort always places an energetic stress on the body and will therefore always lead to an increased utilization of AAs, including the essential ones. In any endurance event this utilization will be maximized when depletion of endogenous CHO pools takes place. Athletes should accordingly know that the breakdown of protein and oxidation of AAs could be limited by the supply of CHO during and immediately after exercise.

MUSCLE PROTEIN

Muscle mass forms the largest protein pool within the body. Muscle protein is thought to be the amino acid supplying pool during starvation conditions (125, 138). Accordingly, starvation is characterized by a decrease in muscle mass and an impaired muscle work capacity. Starvation but also physical exhaustion due to energy deficits is known to change the anabolic/catabolic ratio towards catabolism. As a result, *de novo* synthesis of protein will fall to low levels. Increased degradation and oxidation of protein together with decreased synthesis will then result in a net loss of functional protein, which can be measured as a negative nitrogen balance. Three main goals may be achieved by breakdown of muscle tissue under such circumstances:

1. Liberation of amino acids for use in energy production and maintenance of a normal blood glucose level (gluconeogenesis).
2. Supply of essential amino acids to maintain a normal plasma amino acid composition.
3. Liberation of glutamine to maintain a normal plasma glutamine level, which is assumed to be important for a well functioning immune status and gut function.

Influence of Exercise

Increased AA oxidation as well as nitrogen losses induced by exercise have been described in many studies (25, 64, 106, 107, 157, 187, 188). There has been a debate about whether the AAs oxidized during prolonged exercise stem from the muscle, from the gastrointestinal tract including the liver, or from both. Measurements across particular muscle groups (through determination of arteriovenous AA differences) have shown that some

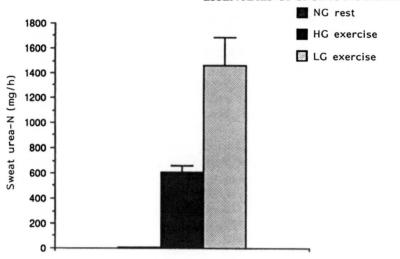

Figure 19 Influence of CHO availability on nitrogen losses in the form of urea-N in sweat at rest and during exercise. NG, normal glycogen; HG, high glycogen; LG, low glycogen. From Lemon and Mullin, *Med. Sci. Sports Exerc.* 1981, 3: 141–149. Reproduced with permission from the American College of Sports Medicine

amino acids are produced by and/or liberated from the muscle during exercise. However, this event may not necessarily reflect a net breakdown of muscle tissue since the muscle is also able to synthesize amino acids. Examples of the latter are alanine from pyruvate and the nitrogen that is derived from the metabolism of BCAAs.

Other events may also influence the protein loss from muscle. Microdamage to muscle fibres due to the influence of mechanical stress may occur (especially) in running events and during negative work (eccentric contractions, e.g. downhill walking or running). Such damage induces repair and inflammation processes after exercise, which will lead to pain perception, most intensively 2–4 days later (so-called delayed onset of muscle soreness, DOMS) (6). Basically these repair processes require the supply of AAs. However, the breakdown of the damaged muscle cells in itself will lead to a liberation of AAs in the same AA pool as that from which AAs are used for *de novo* protein synthesis. Mechanically induced catabolism will thus not necessarily lead to net loss of protein/AAs or to an increased requirement.

VISCERAL PROTEIN

After muscles, visceral tissues (basically abdominal organs) form the second largest protein pool. Amino acids from this pool may be

Figure 20 Former world champion shotputter Werner Günthor. Strength athletes are characterized by a large muscle mass

delivered/exchanged in favour of other pools. The liver, in particular, has been observed to contribute significantly to inter-organ exchange of AAs during fasting and physical stress induced by illness (125).

Influence of Exercise

Exercise may induce an increased contribution of visceral protein to amino acid exchange between organs (157). However, there is some speculation on the quantitative contribution of AAs from this pool to energy metabolism and gluconeogenesis and accordingly to nitrogen loss with sweat and urine during and after exercise. Although it was suggested in the past that the exercise-induced nitrogen loss was derived mainly from muscle protein, there are indications that visceral tissues, which undergo a large reduction in blood flow and in some conditions may become ischaemic (especially the colon (31)), may make a significant contribution (157). Observations from the effects of exercise on intestinal protein turnover indicated a reduced protein synthesis and increased protein degradation during exercise (31, 193). From the previous paragraphs it can be concluded that the main reason for net protein (nitrogen) loss as a result of endurance exercise is the utilization of AAs, derived from different pools, in intermediate energy metabolism. The process is known to be intensified during energetic stress such as a state of rapid energy needs while being glycogen depleted and to lead to a negative nitrogen balance (25, 106, 107, 187, 188).

PROTEIN INTAKE

The average recommended daily intake range for protein in European countries is 54–105 g for adult males and 43–81 g for adult females (183). In comparison, the recommended daily allowance (RDA) in the USA amounts to 58 and 50 g respectively or 0.8–0.9 g/kg body weight/day (131). In general the protein intake in healthy people in the Western world, expressed as en% of total daily intake, amounts to 10–15 en%, resulting in daily intakes of about 50–110 g at energy intakes of 2000–3000 kcal. This figure does not seem to change very much in athletes involved in prolonged heavy exercise. A value of 12 en% was observed during cycling the Tour de France while expending and ingesting 6500 kcal daily over 3 weeks (165). Therefore, it can be concluded that the increased energy intake that is required to compensate for the energy spent in endurance exercise results automatically in an increased protein intake. Accordingly, the latter seems to be a more or less constant percentage of total energy intakes. This relation, however, does not exist in vegetarian athletes, who generally are assumed to have relatively low daily protein intakes, because of a reduced protein density of the foods that they consume (141). Additionally, in general, their energy intake is also relatively low (55). The same is valid for female athletes who consume only low amounts of food in order to maintain a relatively low body weight, such as gymnasts and dancers (176) and, surprisingly, also long distance runners (151).

There is a large body of evidence that the protein requirement of endurance athletes ranges from 1.2 to 1.8 g/kg body weight/day (105–108). There are only limited data on athletes that are involved in strength sports and have a relatively high muscle mass and low fat mass. It is frequently stated that these athletes require more protein than endurance athletes (mainly because of their higher lean body mass) to achieve optimal training status and performance. However, this has often been suggested solely because of their observed high protein intakes, sometimes >4 g/kg body weight (BW) (108). The latter, however, does not mean that this is necessary. Only a few well controlled nitrogen balance studies are available on strength athletes. Tarnopolsky *et al.* (182) determined nitrogen balance in six elite bodybuilders, six elite endurance athletes and six non-training control subjects. He observed that endurance athletes require 1.67 times more daily protein than the non-training control subjects. Bodybuilders needed only 1.05 times more protein to maintain nitrogen balance. However, since this study was carried out over only 10 days and the training load was not explicitly described, it is not clear how well these data represent the true situation during longer periods of daily fluctuating intensive training programmes. Walberg *et al.* (191) studied weightlifters in a weight loss/training regimen. She observed that 0.8 g protein/kg body weight/day resulted in a negative nitrogen balance whereas twice the RDA, 1.6 g kg/day, resulted in a positive nitrogen balance. These data indicate that protein requirements in strength athletes may be only slightly increased. However, athletes with a high energy turnover or consuming low energetic diets may be prone to a decrease in blood glucose levels and glycogen stores, both of which may induce increased AA oxidation. Independent of these nitrogen balance studies it is generally accepted that intakes of 1.5–2.5 g/kg body weight contribute to optimal well-being and performance in strength athletes (108).

A recent study on protein overloading in strength athletes, however, demonstrated that protein turnover more than doubled (both synthesis and breakdown increased by >100%) when 2 g of protein/kg body weight was supplemented, on top of a normal protein intake with food of 1.3 g/kg body weight. As a result nitrogen excretion with urine also more than doubled. During the 4 week training period in this particular study, the strength athletes gained significantly more muscle mass. This suggested that the absolute rate of protein turnover in combination with training stimuli might determine in some way the extent to which lean body mass increases (66). Thus, although intakes in the range of 2.0–3.0 g/kg body weight seem to be superfluous from a requirement point of view, they may have a certain anabolic effect.

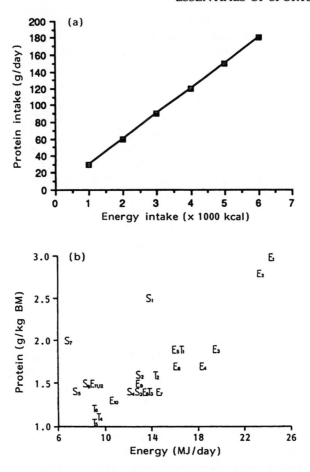

Figure 21 (a) Protein intake expressed as 12 en% of total daily energy intake. Note that an increase in total energy intake will automatically lead to an increased protein consumption. The latter has been observed to amount to about 12 en% in endurance athletes, even during the Tour de France. (b) Protein intake in relation to energy consumption. E, endurance; S, strength; T, team sports. These data show a clear relationship between energy and protein consumption. Athletes who consume less than 1500 kcal may be prone to a marginal or insufficient protein intake. Some protein supplementation to enhance the protein density of the diet may be advised. Reproduced from Erp-Baart *et al.*, *Int J Sports Med* 1989, 10, Suppl. 1: S3-S10, with permission from Georg Thieme Verlag, Stuttgart and New York

PROTEIN SUPPLEMENTATION

In terms of nutritional requirement, it appears that protein supplementation by increasing daily protein intake to a level higher than 12–15 en% will be too high for most athletes. Since a higher daily energy intake in endurance athletes will result in higher protein intake, as well, the value of protein supplementation for endurance sport can be questioned. Based on the observed relationship between energy consumption and protein consumption, athletes expending and eating 5000 kcal/day will ingest twice as much protein as people not involved in exercise and expending/ingesting only 2500 kcal/day. Protein intake for any endurance athlete will thus be sufficient as long as the diet is well composed and contains a variety of protein sources such as lean meat, fish, dairy products, eggs and vegetable protein. Supplementation may be warranted for athletes who compete in weight classes and combine intensive training with weight reduction programmes. Also vegetarian athletes, who consume low energetic and low protein diets (55, 141), or athletes who for any reason are unable to ingest sufficient protein, may benefit from some protein supplementation with the goal to achieve an intake of 1.2–1.8 g/kg body weight a day. Ingesting a moderate amount (10–30 g) of protein powder, e.g. mixed in a liquid, can do this. Examples of supplementation are all categories that are at risk for a marginal nutrient intake as described in Chapter 1 (Table 1), especially those ingesting < 1500 kcal/day (23, 176). Basically the protein sources used for supplementation or as part of replacement meals, taken during 'prolonged endurance exercise days', should be low in fat, easily digestible and of appropriate quality. Milk protein, milk protein hydrolysates and their combinations with whey protein or soya protein are appropriate for these purposes. These protein sources are very low in fat, cholesterol free and do not increase purine intake and uric acid levels in blood. From a health point of view such supplements may also replace a substantial part of the high daily animal protein intake and very frequent egg consumption in high weight strength in order to reduce the atherogenic character of their diets (62).

It should not be overlooked that meals ingested during ultra-endurance events, such as the triathlon, multi-day cycling races and high altitude climbing, and which replace normal meals, may be composed of CHO, fat and protein in a ratio of 60–70 en% CHO, 10–15 en% protein, 25–30 en% fat.

Aspects of amino acid supplementation can be found in Chapter 10. Recent reviews presenting more detail on protein and exercise metabolism can be found in references 224 and 225.

Key points

- Sufficient protein consumption is required for optimal muscle growth and exercise-related repair of muscle damage and enzymatic adaptations.
- The protein requirement of athletes is increased and, according to present knowledge, amounts to approximately 1.2–1.8 g/kg body weight for endurance athletes and about 1.0–1.2 g/kg body weight for strength athletes. The reason for this increase is enhanced utilization of amino acids in oxidative energy production during physical exercise—a process which is known to be intensified at higher endurance work levels and in a state of carbohydrate store depletion.
- There is a close relationship between energy intake and protein consumption. Accordingly, endurance athletes generally ingest a protein quantity that is larger than their required amount. On the contrary, athletes who ingest low caloric diets may also have low protein intakes, which may not compensate for the net nitrogen loss from the body. This may influence protein synthesis processes and training adaptations negatively. To these categories belong bodybuilders, weight class athletes, gymnasts, dancers, female long distance runners and under some circumstances vegetarian athletes.
- Protein intake/supplementation above levels that are normally required will not enhance muscle growth or performance. The building blocks of protein, amino acids, are also involved in numerous metabolic pathways and processes. Some of the amino acids are known to influence hormone secretion and neurotransmission.
- Exercise-induced impairments in neurotransmission are speculated to influence fatigue/performance. However, data that support beneficial effects of single amino acids as present in currently available food supplements are generally lacking.
- The use of single amino acids, to influence metabolic pathways involved in fatigue development and hormone production, needs further research before athletes should be informed positively about benefits (see also Chapter 10).

II Aspects of Dehydration and Rehydration in Sport

5 Fluids and Electrolytes

FLUID RESERVES

Fluid is often forgotten in discussions about nutrient requirements. Humans can live for a prolonged period of time without macro- and micronutrient intake, but not without water. Water is fundamental for all metabolic processes in the human body. It enables transport of substances required for growth and energy production by the circulation and exchange of nutrients and metabolites between organs and the external milieu. Water balance in the body is regulated by hormones and depends on the presence of electrolytes, especially sodium and chloride. The next few paragraphs will explain the importance of water and electrolytes for fluid homeostasis of the exercising individual.

Water is the largest component of the human body, representing 45–70% of total body weight. An average 75 kg human 'contains' about 60% or 45 litres of water. Muscle comprises approximately 70–75% water whereas fat tissue contains only about 10–15% (168). From this it can be deduced that trained athletes who have a high lean body mass and low fat mass have a relatively high water content. Under normal conditions (adequate fluid intake) the body water content is kept remarkably constant. It is not possible to store water in the body, as the kidneys will excrete any excess water. On the other hand it is possible to dehydrate the body by having an imbalance between fluid intake and fluid losses. In such a situation water will be lost from two main compartments in which the water content is normally kept constant.

1. The intracellular compartment.
2. The extracellular compartment.

The extracellular compartment can be further separated into interstitium (space between the cells) and vasculum (space within the blood vessels). A semipermeable cell membrane separates the intracellular water from the water that surrounds the cells. The water content of all compartments is mainly determined by osmotic pressure, caused by osmotically active particles, mainly proteins, electrolytes and glucose. Due to the semipermeability of membranes, as well as ion pumping, the concentration of electrolytes in the intra- and extracellular compartments differs. Water itself can freely pass cell membranes. Osmosis is defined as the passage of water from a region of lower solute concentration to a region with higher concentration. The ultimate result of this water shift is to equalize the two solute concentrations. In the human, body fluid shifts take place to normalize extracellular fluids at an osmolality of approximately 290 mosmol.

Apart from solute concentration, blood pressure also exerts an important effect on fluid exchange. Blood pressure, together with osmotic effects, determines the rate at which water leaves the circulation to enter the tissues, or enters the bloodstream from the tissues. A change in one compartment, e.g. pressure or solute concentration, can directly or indirectly influence the fluid/solute status of the other compartments. For example, during the first few hours of water deprivation, fluid is lost mainly from the extracellular compartment. Blood fluid and plasma volume will decrease, resulting in a compensating water flow from the tissue (interstitium) to the blood. With continuing water deficits the remaining tissue water will therefore become increasingly concentrated. This will initiate water loss from the cells, finally resulting in cellular dehydration. Both extracellular (tissue) and cellular dehydration are known to initiate thirst, a stimulus to ingest water for rehydration (74). Intensive physical exercise, especially when executed in the heat, may lead to dramatic changes in fluid content as well as electrolyte concentration in the different compartments (129, 166–168). Changes in fluid regulatory hormones will stimulate the kidney to reabsorb water and sodium in these circumstances (136). Severe dehydration will initiate impaired metabolism and heat exchange.

INTRACELLULAR FLUIDS AND ELECTROLYTES

Total intracellular fluid content amounts to approximately 30 litres, about two-thirds of the total body water. Water is primarily kept within the cells by an osmotic drive caused by the relatively high electrolyte and protein

Table 2 Approximate concentration (mmol/l) of electrolytes in the intracellular fluid and plasma (114)

	Intracellular (skeletal muscle)	Extracellular plasma water
Cations		
Sodium	10	130–155
Potassium	150	3.2–5.5
Calcium	0	2.1–2.9
Magnesium	15	0.7–1.5
Anions		
Chloride	8	96–110
Bicarbonate	10	23–28
Organic phosphates	65	0.7–1.6

content. An approximate concentration of electrolytes in intracellular fluid is given in Table 2. Sodium and chloride (outside the cells) and magnesium and potassium (inside the cells) are the most important electrolytes exerting an effect on cell water content.

Influence of Exercise

Muscle contractions will result in the production and accumulation of metabolites inside the cell. Initially these metabolites will cause an osmotic gradient leading to a net uptake of water into the cell. At the same time transport processes are initiated and changes in membrane permeability take place. These will lead to transfer of metabolites and potassium from the inner to the outer side of the cell. As a result, the interstitial water will become hypertonic (more concentrated) compared to blood with the result that water will shift from the blood to the interstitium. The synchronically increased blood pressure will further favour this shift (49, 74, 166, 167). As a result, plasma volume will decrease immediately by ±10% after the onset of exercise and will slowly return to a lower level of 3–5% thereafter. Thus, muscle volume increases during exercise as a result of fluid shifts into skeletal muscle. This increase is most pronounced during high intensity anaerobic work, which causes a large intracellular lactic acid production and accumulation.

A secondary haemoconcentration may take place in any case when dehydration occurs during exercise (166). The water pool *between* blood and intracellular space may then be *stressed* from two sides. On the one hand, muscle cells will take up water as described above. On the other hand, large sweat losses will cause plasma volume to decrease and blood electrolyte levels to increase. These changes will draw water from the interstitial space. Finally, if this situation continues, the whole process described initially will be reversed, and intracellular dehydration will take place (74, 166, 168).

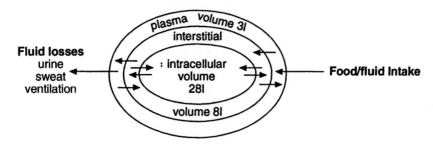

Figure 22 Representation of different water compartments in the body, as well as their fluid exchange routes

EXTRACELLULAR FLUID AND ELECTROLYTES

As described before, the extracellular space can be divided into two subcompartments:

1. Interstitium, the space surrounding the cells and making up the interstitial fluid.
2. Vasculum, the space within blood vessels, for blood plasma.

The total water content of these compartments is approximately 11.5 and 3.5 litres respectively, giving a total of 15 litres extracellular fluid, equal to 50% of intracellular fluid (166). The interstitial fluid is the exchange medium between the cells and the blood. The blood is the final transport medium to deliver oxygen and nutrients to the tissues and to transport water and metabolic end-products such as lactate, ammonia and CO_2 to the lungs, liver, kidneys and skin for elimination and/or excretion.

Regulation of fluid and electrolyte homeostasis, by means of the excretion/retention processes in the kidney, is subject to complex hormonal stimuli (186). The approximate electrolyte concentration of these two subcompartments is given in Table 2. Major differences in electrolyte concentrations exist with potassium and sodium. Potassium is the major intracellular ion. Sodium and chloride are the major extracellular ions. Therefore, sodium and chloride can be regarded as the most important osmotically active electrolytes.

Influence of exercise

The water content of the muscle tissue will increase and blood plasma will decrease, due to repeated muscle contractions. With continuous exercise the water content of all compartments will further decrease as a result of fluid loss by sweating and insensible water loss from the lungs. The latter is normally very small but may be of more impact during activities at high altitude. Metabolic water production during endurance exercise may be significant, but is insufficient to compensate for fluids lost through sweating. Depending on the exercise intensity, training status, climatic circumstances and body size, sweat losses may range from a few hundred millilitres to >2 litres per hour (32, 196).

Because a normal plasma volume is of prime importance to maintain an appropriate blood flow through *exercising tissues*, it may be deduced that a significant decrease in plasma volume will impair blood flow. This will in turn lead to a reduced transport of substrates and oxygen to the muscles that are needed for energy production. Also the transport of metabolic waste products, including heat, from the muscle to the *eliminating organs* such as liver and skin will be impaired. This may lead to a decreased energy

Figure 23 Ultraendurance competitions in the heat may be of risk to health. Minimal clothing in bright colours and regular fluid intake are needed to minimize heat stress

production capacity and fatigue. The decreased heat transfer from the muscles to the skin results in an increased core temperature (32, 114, 167, 168).

In particular, endurance athletes exercising in the heat may be prone to dehydration→heat exhaustion→heatstroke/collapse (129, 167, 168, 181). The electrolyte concentration of sweat is lower than that of blood. This means that relatively more water than electrolytes is lost from the blood. (Sweat electrolyte concentrations are given in Table 3.) Accordingly, dehydration due to sweat loss will lead to an increase in the concentration of blood electrolytes (114).

However, this is only the case when no water is ingested to compensate for the fluids lost. Large sweat losses and compensation by plain water intake may even induce hyponatraemia and consequently signs of water intoxication have been observed in marathon runners and triathletes. A comprehensive review on this topic has been given by Noakes (226). Hyponatremia may exist in symptomatic and asymptomatic forms. The symptomatic hyponatraemia is characterized by a significant decrease in serum sodium, osmolality, plasma volume, intracellular fluid volume as well as extracellular fluid volume. These changes are paralleled by alterations in cerebral function including coma.

Regular endurance training sessions that result in large sweat responses will lead to adaptations in favour of a better maintenance of fluid and electrolyte balance. Sweat glands will adapt to reabsorb sodium and plasma volume tends to increase. Also the sensitivity for fluid regulatory hormones

Table 3 Electrolyte content of whole body wash-down sampled sweat derived from the data of 13 studies

Electrolyte	Cl^-	Na^+	K^+	Ca^{2+}	Mg^{2+}
Average (mmol/l)	28.6	32.7	4.4	1	0.79
SD	13.5	14.7	1.3	0.7	0.6
Average (mg/l)	1014	752	173	40	19
SD	481	339	52	27	15
Range (mg/l)	533–1495	413–1091	121–225	(13–67)	4–34
Bioavailability	100%	100%	100%	30%	35%
Correction factor	0	0	0	×3.33	×2.86
Proposed replacement range (mg/l)	500–1500	400–1100	120–225	45–225	10–100

For the principal electrolytes the table represents 274 observations made on 123 subjects. Net absorption in the gut is assumed to be 100% for Cl^-, Na^+ and K^+ and 30% and 35% for Ca^{2+} and Mg^{2+} respectively. Thus, replacement of the electrolytes lost requires equal amounts of Na^+, Cl^-, and K^+ but larger amounts of Ca^{2+} and Mg^{2+}. Taking a correction for absorption into account reveals an upper replacement level (32, 212). Reproduced with permission from Chapman & Hall, London (142–144). The argument that the sodium content of the meals ingested post-exercise is enough to compensate for the losses is misleading as post-exercise meals do *not* compensate for losses during exercise.

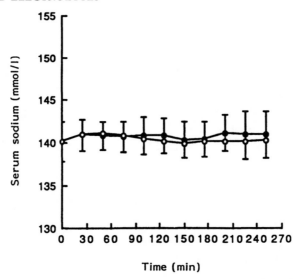

Figure 24 During a 4 hour intensive biathlon (3 hours cycling, 1 hour running) plasma sodium was measured in eight elite triathletes ingesting plain water (— O —) and eight ingesting an isotonic glucose electrolyte drink containing 600 mg sodium/litre (— ● —). Total fluid intake was 600 ml/hour. No effect of sodium intake on serum sodium was observed. Taken from data of Brouns (214)

will be enhanced (114, 129, 196). Sweating will become more 'economical and effective'. Less sweat will drip off the body. Nevertheless, trained people exercising at their maximal levels of endurance performance capacity will be prone to dehydration during competition or intensive training because the thermogenic stress, caused by the extremely high metabolic rates, will initiate maximal sweat rates.

FLUID AND ELECTROLYTE INTAKE

Daily fluid intake is normally associated with food consumption (salty/spicy foods) and with having a dry mouth. To a large extent this accounts for learned (conditioned) drinking behaviour. True thirst, however, arises as a consequence of intra- and extracellular dehydration (74).

In general, fluid intake should equal total daily water turnover, which is assumed to be about 4% of body weight in adults (131). Total water turnover can vary markedly, mainly because of differences in metabolic rate (exercise will greatly influence this factor) and in insensible water loss. The latter can be strongly influenced by climatological circumstances as well as by altitude. Acute water loss in large quantities can also result from diarrhoea. The daily water requirement basically represents the amount

Figure 25 Fluid post in the 67 km Swiss Alpine Marathon of Davos. Fluid and carbohydrate are required to maintain optimal performance

necessary to balance insensible losses (via breathing and skin) and to supply the kidneys with the minimal amount of fluid needed for excretion of metabolic end-products, such as urea, and electrolytes. A minimum fluid intake of 1.5–2.0 l/day for a 70 kg male may be needed to avoid metabolic disturbances and kidney problems.

For normal sedentary individuals, an intake level of 1 ml/kcal energy expenditure seems a general recommendation (131). A normal fluid intake in line with a normal daily water turnover amounts accordingly 2.5–3.0 l/day. This principle may, at least in some conditions, also apply to athletes who have a higher energy turnover. Cycling a mountain race for example, while expending 6000 kcal/day may then require at least 6 litres of fluid. A level of 6 litres fluid intake has been reported under these circumstances (165). Running a marathon (energy cost approximately 3000 kcal (137)) would then cause an extra fluid requirement of 3 litres.

The *minimal* daily requirements for adults given by the National Research Council (1989) for sodium, chloride and potassium, the major electrolytes active in water homeostasis and also lost by sweating, are 500, 750 and 2000 mg respectively. Daily food intake generally leads to much higher intakes than these figures. Therefore, supplementation is not advisable. However, in the case of substantial losses such as during acute diarrhoea or as a result of prolonged intensive sweating, electrolyte levels in plasma may be threatened. In these cases it is advisable to include some electrolytes in rehydration solutions.

REHYDRATION SOLUTIONS

Rehydration solutions for athletes are generally designed to replace fluid and minerals lost by sweating and also limited amounts of energy in the form of CHO. All three substances are either lost or used during endurance exercise. Higher exercise intensities require a higher degree of energy production for which CHO as energy source is most suitable. Accordingly, with higher exercise intensities, more metabolic heat will be produced. Consequently sweat rate will be increased, as will the excretion of electrolytes. The longer the exercise lasts, the larger the amount of fluid, electrolytes and CHO needed to replace the losses.

There are large differences between individuals in sweat rate, sweat electrolyte content, degree of CHO utilization, etc. These differences can be further influenced by climatological circumstances. As a result, it is impossible to recommend a general rehydration solution that will exactly compensate for the losses of any individual in any situation. Commercial rehydration solutions are generally designed to cover the needs of a large exercising population under different circumstances. This is necessarily a compromise that has to be made by any producer.

General guidelines for the composition of rehydration solutions have been obtained from a large number of studies in the field of gastric emptying, intestinal absorption, fluid balance regulatory factors and fatigue/performance and have been summarized in a number of excellent reviews (30, 32, 40, 47, 69, 113, 114, 126, 130, 153, 155). The general outcome from these studies is that addition of small to moderate amounts of CHO to a drink does not delay gastric emptying and improves absorption, compared to plain water. The scientific rationale behind these findings is the fact that coupled glucose–sodium transport across the gut membrane is very fast and stimulates water absorption due to the osmotic action of these solutes when being absorbed (69, 114, 131).

The addition of electrolytes, in small quantities as lost by the whole body sweat, will influence neither gastric emptying nor absorption (153, 154). The CHO fraction will contribute to the maintenance of a normal blood glucose level and will lead to a sparing of the endogenous CHO reserves (49, 75,

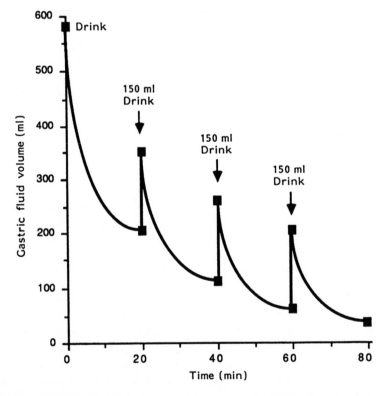

Figure 26 Gastric emptying rate after ingestion of a single bolus (600 ml) of an isotonic carbohydrate (7%)-electrolyte solution or with repeated drinking as usual in endurance events. Repeated drinking with 70 g CHO/litre does not lead to fluid accumulation in the stomach. Taken from data of Rehrer *et al.* (215)

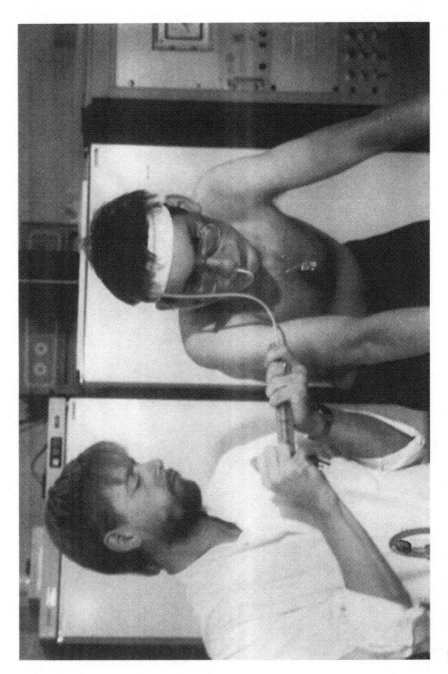

Figure 27 Determination of gastric volume and gastric secretion by using a naso-gastric tube during cycling exercise

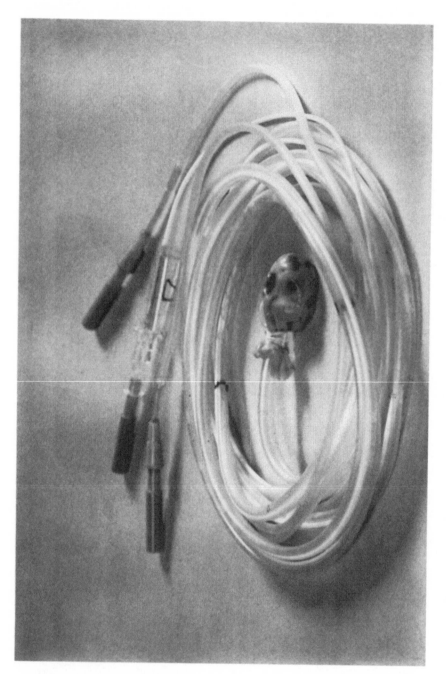

Figure 28   A triple lumen catheter, which is used to study fluid and substrate fluxes in the jejunum

124, 148). The latter may influence protein breakdown, delay fatigue and thus influence performance (25, 43–45, 113, 127, 188, 189).

A large body of scientific evidence shows that different types of CHO in amounts of 30–80 g/l and sodium in amounts of 400–1100 mg/l induce a high rate of gastric emptying and fluid absorption (69, 114). A maximal fluid absorption rate seems to be a prerequisite only in the event that the quantity of fluid ingested balances or exceeds the quantity that can be absorbed at the same time (for example in the case of massive fluid loss in watery diarrhoea). However, fluid intake during exercise generally does not exceed 600 ml/h in runners or 800 ml/h in cyclists (143). This seems less than the amount that could be absorbed maximally. Therefore, it is still open for discussion whether a maximal rate of gastric emptying and absorption is always necessary for the exercising individual. Thus, slightly more concentrated CHO–electrolyte solutions (up to 100 g CHO per litre) are known to reduce the rate of net fluid absorption, but enhance CHO availability. In the case of submaximal fluid intake, more concentrated

Table 4 Oral rehydration solutions for combined fluid carbohydrate–electrolyte supply in sports

Recommended		Optional	
Carbohydrate	30–100 g/l[b]	Chloride[a]	max. 1500 mg/l
Sodium[a]	max. 1100 mg/l	Potassium[a]	max. 225 mg/l
Osmolality	<500 mosmol/l[c]	Magnesium[a]	max. 100 mg/l
	favourable ≤ isotonicity		max. 225 mg/l

Carbohydrate sources:	Maximal amount of CHO (to avoid hypertonicity and/or a too high concentration[b])
Fructose	35 g[d]
Glucose	55 g
Sucrose	100 g
Maltose	100 g
Maltodextrins	100 g
Dispersible starch	100 g

[a] Quantities taken from Table 3.
[b] Water absorption becomes maximized with approximately 30 g CHO/litre. This is also about the minimum amount of CHO needed to achieve measurable effects on glucose/energy metabolism. The upper level (100 g) is given because gastric emptying rates and therefore fluid availability will be reduced too much at higher concentrations. Additionally the osmotic load of drinks containing more than 100 g will be increasingly effective in reducing the net fluid absorption. More concentrated solutions cannot be considered as rehydration drinks, but are energy (CHO) supplements.
[c] Net water absorption in the gut, after gastric emptying, is mainly determined by substrate absorption—which pulls water along, and by osmotic gradients. An increase in solute (carbohydrate) concentration will lead to a higher solute absorption and with it water absorption. An increase in osmotic load, however, enhances osmotic fluid secretion into the gut. Net fluid absorption results from two opposite water fluxes (absorption–secretion). Thus, hyperosmolality will counterbalance water absorption benefits achieved by solute transport. Osmolalities of >500 mosmol/l should be avoided.
[d] Fructose as sole CHO source may induce gastrointestinal distress at concentrations of >35 g/l. This is not the case in combination with other CHO (e.g. sucrose).

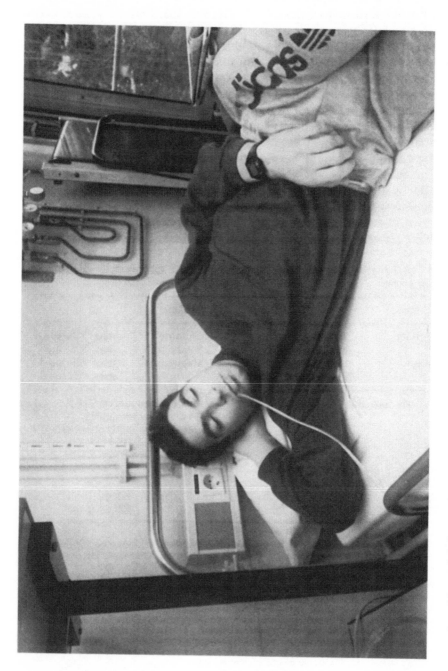

Figure 29 An athlete while being perfused through the triple lumen catheter. These complex perfusion experiments generally last 8–10 hours

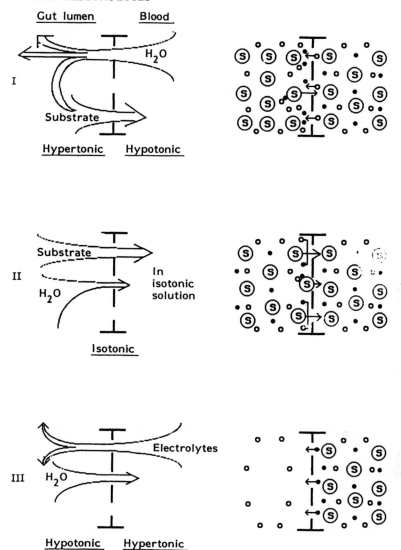

Figure 30 A schematic representation of osmotic effects in the gut. Plain water perfusion will induce electrolyte secretion and water–electrolyte absorption. Hypertonic perfusion will induce water secretion and water–substrate absorption. Isotonic perfusion will induce substrate–water absorption. (Net absorption = absorption − secretion.) O, water; ●, electrolyte; S, solute

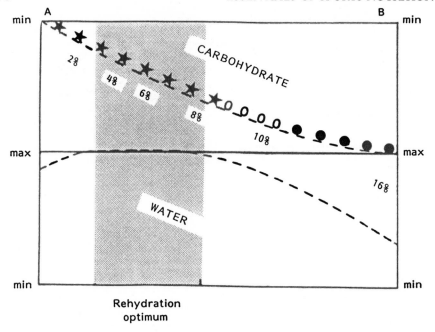

Figure 31 Low amounts of CHO stimulate water absorption (left part of figure, 'A'). High amounts of CHO in a beverage reduce gastric emptying and induce fluid secretion, leading to a reduced net fluid absorption (right part, 'B'). 'A' leads to high fluid-low CHO availability. 'B' induces high CHO–low fluid availability. Maximal CHO availability, without impairing fluid homeostasis is found with beverages containing 60–80 g of CHO/litre. Optimal choice of a drink depends on climatological circumstances and physiological characteristics of the sports event. Reproduced from Brouns (32) with permission from Chapman & Hall, London

drinks have similar effects on fluid homeostasis to water or very dilute CHO solutions (32, 40, 114, 126, 127).

Flavoured drinks are preferred by athletes compared to plain water. Consequently such drinks are ingested in larger volumes (88). A general guideline should be that rehydration solutions should not be strongly hypertonic (i.e. <500 mosmol and preferably <300 mosmol). Drinks in the low hypertonic range (414 mOsm) do not differ significantly in rate of fluid absorption, urine production and plasma volume, from isotonic (297 mOsm) or hypotonic drinks (197 mOsm) (491). Hypertonic solutions have been shown to reduce the rate of net fluid absorption by inducing fluid secretion into the gastrointestinal tract. Additionally, they may also reduce the rate of gastric emptying. The latter may lead to feelings of fullness and influence/limit quantitative fluid consumption (30, 32, 114, 115, 155).

Table 5 Carbohydrate content and osmolality of selected drinks

	CHO (g/l)	mosmol/kg
Sport energy drinks		
Dextro Energy Fruit	110	956*
Exceed	72	270
Extran Orange	145	959*
Isostar Long Energy	152	303
Leppin enduro booster	97	137
Perform Energy drink	165	397
Rehydration drinks		
AA Drink	68	330
Aquarius	63	400
Athlon	62	274
Enervit Tropical	70	320
Extran Citron	75	332
Rivella activ	9.6**	96
Gladiators	90	392
Lucozade low cal	6.0**	148
Gatorade	60	378
Isostar	70	281
XL-1	61	291
Soft drinks and 'designer' drinks		
Apple juice	104	695
Coca-Cola	105	650
Fanta	108	478
Orange juice	94	662
Sprite	110	591
Red Bull***	107	686
Taurus***	S119	795
Guarana Jones***	100	613
Flying Horse***	107	862

CHO content as labelled on the product. Osmolality was measured by using a freezing point depression osmometer (227).
* An osmolality of this magnitude may result in gastrointestinal distress during exercise. It is recommended to dilute these drinks by >100% to obtain an osmolality of <400.
** The carbohydrate content is too low to result in significant energy support during exercise.
*** Caffeine content of 320 mg/l.

The source of CHO will influence fluid osmolality. Therefore, so as not to result in very high osmolalities, the quantity of monosaccharides dissolved is recommended to be smaller than that of disaccharides or polysaccharides. Based on current knowledge and evidence, a general recommendation for the composition of oral rehydration beverages for sport is given in Table 4. Table 5 gives examples of the composition of commercially available sport drinks and other drinks.

Key points

- Fluids and electrolytes are important for the maintenance of fluid balance during prolonged physical exercise, especially in the heat.
- Progressive fluid loss from the body, by means of sweating and breathing, is associated with a decreased blood volume and blood flow through the extremities. Also a reduction in sweating and heat dissipation may result from this. Under circumstances of high intensity work in the heat, it may lead to heat stroke and collapse.
- Dehydration of >1.5 litres is known to reduce the oxygen transport capacity of the body and to induce fatigue and gastrointestinal disturbances.
- Appropriate rehydration is known to counter these effects and to delay fatigue. In contrast to plain water, the addition of CHO to rehydration drinks is known to stimulate drinking and water absorption and additionally to have a positive effect on water balance.
- The carbohydrate supplied with the drink will also be of benefit for maintaining a high CHO availability, to help reduce fatigue and maintain performance capacity.
- Addition of sodium to drinks will have a positive effect on post-exercise rehydration by reducing urine loss and stimulating water retention. Other electrolytes may be added but should not exceed the levels of loss with whole body sweat; they have not been shown to have a beneficial effect on performance.
- Sport rehydration drinks should in principle not be hypertonic.

III Nutritional Aspects of Micronutrients in Sport

6 Minerals

Minerals are essential for a well functioning skeleton and musculature. Growth requires minerals as building substances and an insufficient supply of calcium and phosphate is associated with impaired skeletal development. Minerals are important for nervous transmission processes, muscle contraction, enzyme activity, etc. In the previous chapter on fluids and electrolytes, the role of sodium and chloride in fluid homeostasis was described. Here we shall briefly describe how minerals other than sodium and chloride are involved in important biological actions of the exercising individual and how their requirements are influenced by exercise.

The minerals to be discussed are:

- Potassium
- Magnesium
- Calcium
- Phosphorus
- Iron
- Zinc

MINERAL RESERVES

The mineral content in the body differs among tissues as well as between intra- and extracellular compartments. Bone has a high calcium and phosphate content, the muscle cell has a high content of potassium and magnesium, and blood and interstitial water are high in sodium and chloride. Although minerals are fixed components of tissues such as bone or muscle, this does not necessarily mean that they are freely available for metabolic purposes. The major fraction of the 'metabolic' mineral pool is present in blood plasma and interstitial fluid.

The amount of minerals circulating in body fluids is a resultant of different ongoing processes. Absorption from food on the one hand and uptake or release by tissues as well as losses/excretions (by sweat, urine, faeces) on the other hand, determine the actual mineral content. This mineral level remains within a narrow range. Therefore, any excess of minerals will be compensated by increased excretion. Any shortage will, in the first instance, be compensated by reduced excretion or/and by increased release from tissues. With a continuing shortage, plasma mineral levels will start to fall. The latter will influence the uptake or release of minerals by the cells and thus the cell mineral status. During prolonged periods of mineral deficits, cell growth and cell function will become impaired. However, a short period with a relative shortage of one or more minerals, e.g. as may occur during an ultra-endurance event, will not necessarily mean that health and performance are affected.

POTASSIUM

Potassium is the major intracellular cation with a concentration of about 40 times the concentration of extracellular water (see Table 2). Potassium is important for the transmission of nerve impulses, membrane potential and hence muscle cell contraction, and maintenance of normal blood pressure. Most (90–100%) ingested potassium is absorbed in the gut and enters the circulation (131). The plasma potassium content has been shown to influence the contractility of both heart and skeletal muscle. Excessive plasma potassium levels produce typical changes in the electrocardiogram and may even lead to a sudden heart standstill. Therefore, large intakes of potassium, leading to excessive blood potassium levels, should be discouraged (131). Potassium is excreted from the body in urine and to a small degree in faeces and sweat. Diarrhoea is known to result in high potassium losses.

Influence of Exercise

Potassium is lost from the muscle cells during repeated contractions. This loss is caused by changes in cell permeability and the frequent inward and outward fluxes of sodium and potassium that are part of the electro-chemical contraction process (121, 185). In muscle cells potassium is stored within glycogen (18). Accordingly, breakdown of glycogen will lead to a liberation of potassium in the muscle cell and may subsequently enhance potassium loss from the cell into the extracellular space. As a result, the potassium concentration in interstitial fluid, as well as in blood plasma, will increase. This increase will be most pronounced during maximal exercise intensity (121, 185). It has been suggested that potassium may also be lost from damaged muscle fibres, but no evidence for such loss is available. Muscle fibre damage occurs due to mechanical stress, primarily during activities involving negative work, such as downhill walking/running (6). Sweat losses incurred during exercise will result in only small potassium losses. The concentration of potassium in sweat is about equal to that in blood plasma. Post-exercise, potassium is excreted in larger quantities in the urine, most probably because the kidney is stimulated to retain sodium for fluid homeostasis and will therefore exchange sodium for potassium (30, 114). Some concern has arisen in the past about the possible effect of sustained exercise and sweat losses, during ultra-endurance events, on the plasma potassium concentration as well as potassium balance. It has been thought that prolonged exercise-induced losses may lower plasma potassium to such an extent that muscle and heart function will be compromised. However, since prolonged exercise leads to a continuous efflux of potassium from the muscle, plasma potassium has not been shown to fall. Any deficit will therefore occur in the intracellular potassium levels that are difficult to measure. However, intracellular potassium losses may be more than compensated by the release of potassium from the breakdown of intracellular glycogen. In this case there would be no change in intracellular free potassium. Increased potassium requirements *during exercise* are, therefore, unlikely.

Post-exercise potassium requirements may be enhanced. Immediately after exercise there is a very rapid uptake of potassium by the muscle. Muscle glycogen synthesis and coupled potassium uptake proceed at a high rate. As a result, plasma potassium levels are known to decline very rapidly after the finish of exercise to normal resting levels or slightly below (121, 185).

Potassium Intake

The recommended minimal daily intake for potassium is 2 g/day (131). This figure does not take into account the possible exercise-induced losses through sweat and urine. The desirable intake, therefore, is 2–3.5 g/day (52, 131). Potassium is widely available in foods as it is an essential constituent of all living cells, especially fruits (bananas, oranges), vegetables (potatoes) and meat. Accordingly, the potassium intake may vary considerably depending on food selection. High intakes of particular food items may lead to a potassium intake as high as 8 g/day (131).

MAGNESIUM

The magnesium content of the body is approximately 20–30 g. About 40% of this amount is located within the cells (especially muscle), about 60% in the skeleton and only 1% in extracellular fluid (155). Magnesium is an essential mineral present in about 300 enzymes that are necessary for biosynthetic processes and energy metabolism.

Magnesium plays an important role in neuromuscular transmission and activity: it acts at some points synergistically with calcium, while at others it is antagonistic. As with all minerals, the magnesium level in blood plasma is kept within a narrow range. Practically all metabolically available magnesium is within the very small extracellular pool (see Table 2). Any change in this pool is caused by nutritional intake, by uptake in or release from tissues or by losses or excretion (37, 114). The fractional magnesium absorption in the gut is approximately 35%. Magnesium is excreted mainly

Figure 32 Tropical fruits and tomatoes as well as their juices have a high potassium and antioxidants content

in the urine and small amounts are lost with sweat (see Table 3). Faeces also contain magnesium but this represents the unabsorbed magnesium fraction.

Influence of Exercise

Low resting and exercise plasma magnesium levels have repeatedly been reported in athletes involved in regular endurance exercise. This has been thought to lead to impaired energy metabolism, greater fatigue and the occurrence of muscle cramps (37, 131), although the latter could not be confirmed in a study on marathon runners (211).

Several different explanations have been given for this decrease. It has been suggested that it results from magnesium loss through sweat as well as from an enhanced uptake by red blood cells and fat cells. Therefore, it is difficult to decide whether a reduced plasma magnesium level in athletes represents a true marginal (deficit) status or whether this is simply a result of physiological magnesium shifts. Terblanche et al. (228) studied the effect of Mg supplementation on muscle magnesium content and running performance during a marathon as well as on running induced muscle damage and post-exercise muscle recovery. Twenty experienced marathon runners ingested 365 mg magnesium for a period of 6 weeks. This supplementation did not affect any of the parameters studied. Maximum voluntary muscle contraction was significantly decreased after the marathon, and 7 days of recovery were required for this parameter to return to pre-race control levels. Magnesium supplementation did not influence this recovery process. It was concluded that magnesium supplementation in athletes that are healthy and do not have magnesium deficits is not beneficial. Similar findings were reported by Weller (483). The losses through sweat are generally small (see Table 3) but may become significant with prolonged high sweat rates. Additionally, magnesium loss may be increased during the first 24 hours after a strenuous exercise.

Magnesium Intake

The recommended daily intakes for minerals other than sodium, potassium and chloride are given in Table 6. The data given in this table represent the quantities established by expert panels in the USA and Germany. These quantities are thought to be adequate for sedentary people. As yet there are no guidelines for athletes, who may have higher daily requirements for most nutrients.

The magnesium content of food varies widely. Fish, meat and milk are relatively poor in magnesium, while vegetables, exotic fruit, berries, bananas, mushrooms, nuts, legumes and grains are relatively rich.

Magnesium intake has been found to decline over recent decades, most probably due to the increased consumption of refined and processed foods.

Table 6   Recommended dietary allowances for minerals (mg)

Age	Magnesium	Calcium	Phosphorus	Iron	Zinc
Males					
15–18/15–18	400/400	1200/1200	1200/1600	12/12	15/15
19–24/19–25	350/350	1200/1000	1200/1500	10/10	15/15
25–50/25–51	350/350	800/900	800/1400	10/10	10/15
Females					
15–18/15–18	300/350	1200/1200	1200/1600	15/15	12/12
19–24/19–25	280/300	1000/1000	1200/1500	15/15	12/12
25–50/25–51	280/300	800/900	800/1400	15/15	12/12

The data given in milligram (mg) are derived from NRC/DGE.
NRC = National Research Council, Recommended Dietary Allowances, 1989 (USA).
DGE = Deutsche Gesellschaft für Ernährung, Empfehlungen für die Nährstoffzufuhr, 1991.

For example, > 80% of the magnesium found in whole grain is lost by removal of the germ and outer layers (131). Data concerning the magnesium intake in athletes are scarce. A Dutch study, however, indicated a close relationship between magnesium intake and energy intake (59). Endurance athletes in particular, having higher daily energy intakes, were observed to have an adequate daily magnesium supply compared to the daily recommended allowance for sedentary people. The latter, however, may not represent optimal quantities for athletes, since magnesium losses with urine and sweat are increased as a result of intensive training and are not taken into consideration in the RDA.

CALCIUM

The human body contains about 1200 g of calcium of which approximately 99% is fixed in the skeleton. Only a fraction (1%) is present in extracellular fluid and intracellular structures of the soft tissues (131). This small fraction represents the metabolically available pool. Plasma calcium is maintained in a narrow range mainly by hormones that control absorption, secretion and bone turnover. Calcium entering the plasma is derived from food or from release from bone tissue. Calcium is lost through urine, sweat and faeces. The calcium present in faeces mainly represents unabsorbed calcium. In adults the fractional calcium absorption in the gut amounts to approximately 30% (131). Urinary excretion is largely influenced by food intake. Urinary calcium excretion has been observed to increase with higher protein intake levels, especially if phosphorus intake remains at the same level (110, 131). Bone is constantly turning over, thus constantly absorbing and releasing calcium together with phosphate. When calcium intake is too low, plasma calcium levels will remain constant due to an enhanced release from bone.

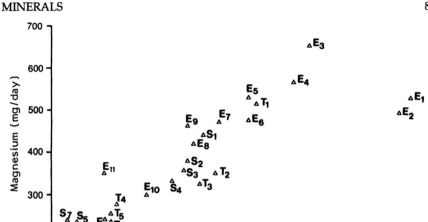

Figure 33 Magnesium intake in athletes increases with higher energy intakes. E, endurance athletes; S, strength athletes; T, teamsports athletes. Reproduced from Erp-Baart *et al.* (59) with permission from Georg Thieme Verlag, Stuttgart, New York and Erp-Baart (60)

Influence of Exercise

During exercise, calcium plays an essential role in initiating muscle contraction. Calcium liberation within the cell initiates a state of contraction whereas re-uptake initiates relaxation. Plasma calcium has been shown to remain unchanged, to decrease or to increase during exercise (37, 169). This variation may be attributed to different factors such as water loss leading to concentration, an increased release from bone due to mechanical stress or an a reduced uptake by bone due to decreased bone mineralization.

A large body of scientific evidence has recently shown that female athletes may suffer from stress fractures and/or reduced bone density. This 'athletic osteoporosis' has been associated with depressed oestrogen levels due to exercise stress. Oestrogen is known to regulate calcium metabolism. Additionally, the relatively low calcium intakes in these female athletes may have a large impact (37). This causes a high frequency of this abnormality among female athletes. Athletes involved in strength training and consuming a high protein diet may excrete more calcium through the urine, especially when phosphorus intake is not increased in parallel to protein intake. Calcium losses through sweat are very small (see Table 3).

Calcium Intake

Calcium intake also varies widely according to the quantity and composition of the diet. Dairy products are a major source of calcium intake. Nuts, pulses, some green vegetables (broccoli) and seafoods as well as calcium from drinking water may further contribute. Daily calcium intake depends both on food selection and the total food/energy intake.

Athletes with low daily energy intake or those who follow a weight reduction programme may therefore have a marginal calcium intake. Females, especially long distance runners, have often been found to have calcium intakes that are lower than the RDA, probably as a result of relatively low energy intakes (37, 59, 84, 131, 206). It has been reported (87) that a calcium intake of 1500 mg/day is required to achieve calcium balance in postmenopausal women not receiving oestrogen replacement therapy. Barr (9) concluded from these data that female athletes who are amenorrhoeic and have low oestrogen levels should ingest 1500 mg

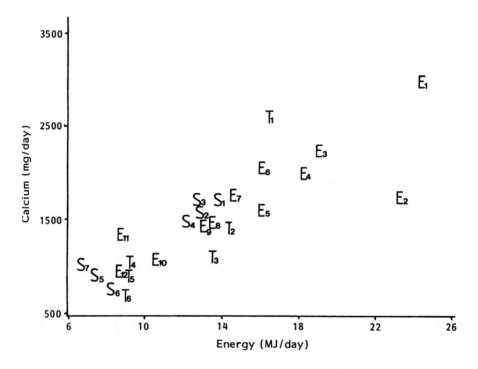

Figure 34 Calcium intake in athletes increases with higher energy intakes. E, endurance athletes; S, strength athletes; T, teamsports athletes. Reproduced from Erp-Baart *et al.* (59), with permission from Georg Thieme Verlag, Stuttgart and New York

calcium/day. On this basis, all amenorrhoeic athletes (at risk groups include runners, dancers, gymnasts, bodybuilders) would have inadequate intakes.

PHOSPHATE

Phosphate is the counterpart of calcium in bone formation. About 85% of the total phosphate is present in the skeleton. The remainder is distributed between extracellular and intracellular space in soft tissue. Phosphate is an essential element in numerous enzymes as well as in energy metabolism (nucleotides and conjunction with B vitamins). Phosphate intake, and consequently supply to the blood, is known to affect bone formation. Therefore, the intake of phosphate and calcium should be balanced. The fractional phosphate absorption in the gut is approximately 70%, which is about twice as high as that of calcium absorption (131). Phosphate is mainly excreted in the urine, the unabsorbed fraction in the intestine leaves the body with the faeces and minor amounts are lost with sweat.

Influence of Exercise

Exercise that leads to a substantial sweat loss results in haemoconcentration, which in turn will elevate plasma phosphate levels. Phosphate losses through sweat are negligible. In addition, changes in alkalosis (inducing a fall in phosphate levels), acidosis and cell damage (inducing an increase in phosphate levels) are known to influence plasma levels (104).

Phosphate Intake

Phosphate is especially present in protein-rich foods such as milk, meat, poultry and fish, as well as in cereal products. The amount of phosphate that is present in the normal diet is about 1500 mg. This figure has been relatively constant over the years. Increases in daily energy intake will normally also lead to an increased phosphate intake. Therefore, phosphate deficiencies normally do not occur in healthy exercising individuals (131).

IRON

Iron is an important constituent of haemoglobin, myoglobin and a number of enzymes. As such, the availability of iron is important for oxygen binding capacity of the red blood cells, and transport as well as for the transfer of electrons in the electron transport chain. About 30% of total iron is found in storage forms as ferritin and haemosiderin and a small part as transferrin. Therefore, these iron stores can serve as indicators of iron status. Poor iron status may be indicated by low levels of serum ferritin, increased red cell protoporphyrin levels, reduced transferrin saturation levels and reduced haemoglobin levels. With inadequate iron intake the storage form will be the first to be affected. With prolonged iron shortage haemoglobin production will finally be affected, resulting in iron deficiency anaemia. The latter will reduce oxygen transport capacity and may thus affect endurance performance capacity (131, 145, 150).

Influence of Exercise

There is considerable controversy about the extent to which athletes are any more iron deficient than the normal population, especially with respect to haemoglobin concentrations, which are known to be relatively low in many endurance athletes. An explanation for this observation may be the plasma volume increase as a result of endurance training. The absolute amount of circulating haemoglobin in this case is not necessarily lower but rather an effect of the increased plasma volume resulting in pseudo anaemia. However, over the last 10 years a large body of evidence has indicated that a substantial number of athletes involved in regular training do also have decreased iron stores, indicated by reduced bone marrow iron, enhanced iron binding capacity and low serum ferritin levels. Serum ferritin levels have to be considered with care as stressful exercise has been shown to result in temporarily increased levels. Thus serum ferritin levels obtained shortly after intense endurance exercise may not accurately reflect body iron stores (102).

There is some evidence that the poor iron status observed in athletes can be partly explained by consumption of a diet that is poor in (biologically bound) haem-iron. This seems to be especially the case in athletes who consume vegetarian and high fibre meals. While a relatively small amount of iron is absorbed from the diet, a significant amount of iron may be lost with sweat. This may explain the effects of intense endurance training on iron status. Yet, there are a number of other hypotheses that have been put forward to explain the low iron stores observed in athletes. Mechanical stress in the footsole during the landing phase of the foot while running has been suggested to lead to red blood cell damage. The latter is assumed to lead to haemolysis and loss of haemoglobin. However, from a mechanistic point of view, the iron from the haemoglobin will re-enter the circulating iron pool and will thus be biologically available again (37, 56, 57, 118, 132, 145, 150, 152). Another hypothesis concerns the effect of exercise on iron absorption in the gut, especially ultra-endurance activities that lead to substantial sweat losses which reduce intestinal blood flow and may lead to damage of the gut epithelium resulting in blood loss. As a result, increased faecal haemoglobin and iron loss have been observed (30, 123).

Iron Intake

Red meat, liver, poultry, dark green vegetables and cereals (especially iron fortified products) are the major sources of iron intake. Haem-iron in meat is the best absorbable iron source. The food matrix, i.e. the way a meal is composed, can influence iron absorption. Vitamin C enhances inorganic iron absorption, while components in dietary fibre, tea, coffee and phosphate reduce absorption (131).

As described above, iron intake in vegetarian athletes is often low. Female athletes or athletes who compete in weight class sports or gymnastics also may have inadequate iron intakes as a result of the consumption of low calorie diets (38, 55, 84, 176, 206). The recommended/safe daily intake for iron is shown in Table 6. It is assumed that the required daily intake for athletes exceeds this RDA. Research is needed to establish the real athlete's requirements.

ZINC

Zinc is present in relatively large amounts in bone and muscle. However, as is the case with other minerals, these stores are not metabolically available. The pool of zinc that is readily available circulates in blood, is small and has a rapid turnover rate. Zinc is involved in growth and development of tissues, especially muscle, as it is an essential substance in numerous enzymes involved in major metabolic pathways. Recent studies have indicated that zinc may also play a crucial role in immune competence (5, 97, 131).

Influence of Exercise

Serum zinc represents to a large extent the metabolically available zinc pool. Any rapid change in blood volume caused by physical exercise will affect serum zinc status either by dehydration, which will increase zinc concentration due to haemoconcentration, or there will be a post-exercise plasma volume increase caused by water and sodium retention (this will decrease zinc concentration). Apart from these effects it is assumed that functional zinc shifts between tissues may occur during exercise. It is therefore difficult to determine the effect of exercise on zinc status as indicated by serum zinc levels only (5, 37). For example, plasma zinc levels increased significantly over several weeks in cycling participants in the Tour de France, although a decrease was expected (164).

Nevertheless, regular exercise may increase zinc requirements because zinc from the body is primarily lost by urine and sweat and both are enhanced as a result of endurance exercise (4, 5, 37, 41, 78). Zinc loss with sweat may be substantial. However, large individual variations do exist (230). When ingesting a normal diet, repeated days of prolonged running in the heat did not cause a decline in plasma zinc levels (231).

Figure 35 The footsole is prone to high pressure peaks during the landing phase while running. It is hypothesized that red blood cells with a low stress tolerance may be damaged, leading to anaemia

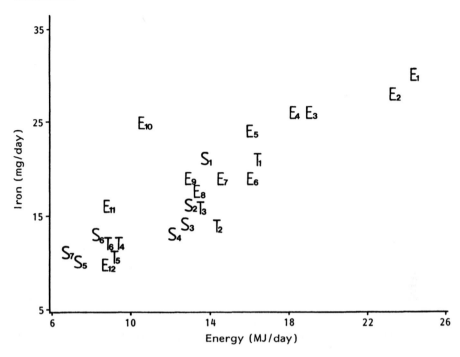

Figure 36 Iron intake in athletes increases with higher energy intakes. E, endurance athletes; S, strength athletes; T, teamsports athletes. Reproduced from Erp-Baart *et al.* (59), with permission from Georg Thieme Verlag, Stuttgart and New York

From this it may be concluded that a decline in serum or erythrocyte zinc levels as has been observed in some studies (232–235) may be due to a combination of low zinc intake and exercise induced zinc losses. Long-term fasting, leading to a negative nitrogen balance, may induce muscle loss and has been reported to increase serum and consequently urine zinc levels (178). Similarly, it has been thought that in the post-exercise phase, zinc may be lost from damaged cells that are broken down in repair processes. A recent study, by Kazunori and Clarckson (216), however, did not confirm effects of muscle damage on serum zinc levels. Also Nosaka (229) studied the effects of eccentric exercise on muscle damage and plasma zinc changes. Although the exercise resulted in muscle damage and soreness, there was no effect on plasma zinc levels.

Zinc Intake

Meat, liver and seafood are major zinc sources in the diet. Additional zinc may be derived from milk and cereal products. High CHO foods, especially

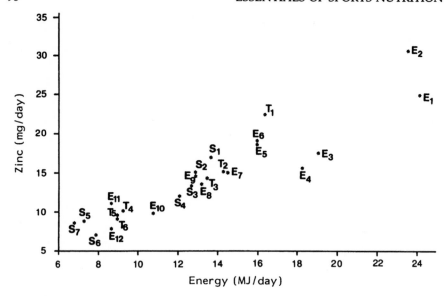

Figure 37 Zinc intake in athletes increases with higher energy intakes. E, endurance athletes; S, strength athletes; T, teamsports athletes. Reproduced from Erp-Baart *et al.* (59) with permission from Georg Thieme Verlag, Stuttgart and New York and Erp-Baart (60)

refined sources, are poor zinc sources. Phytate and dietary fibre are known to reduce zinc absorption (131) and may thus reduce the bioavailability of zinc when present in the diet in significant amounts. Research shows that the daily dietary zinc intake appears to be marginal in many sedentary individuals. The effect of low zinc intakes on zinc status may be exacerbated in athletes by elevated exercise-induced losses (84, 177). However, zinc intake in athletes is closely related to the total energy intake. Accordingly, most elite athletes studied in The Netherlands have higher intakes than RDA for normal sedentary people (60). Vegetarian athletes may have low zinc intakes (72, 141). However, there is no good evidence that vegetarian persons, in general, are zinc deficient (55). The recommended/safe daily intake for zinc is given in Table 6.

MINERAL REPLACEMENT AND SUPPLEMENTATION

From the previous paragraphs it may be concluded that mineral intake in athletes, compared to RDA for normal sedentary people, is sufficient in most cases. Further, supplementation in healthy individuals, consuming a well balanced diet containing sufficient amounts of meat, fruit, vegetables, cereals as well as whole grain products, may not be beneficial. But, for a number of reasons, the diet of athletes involved in intensive training is often unbalanced. Many athletes consume up to 40% of daily energy intake as in-between snacks that are rich in energy, but poor in micronutrients (23, 58, 165). Mineral intake largely depends on both the quality of the food selected as well as the quantitative food intake. Thus, only in cases of low energy diets or in periods of inappropriate food intake due to appetite loss when not feeling well is there a reason for supplementation. The addition of minerals to products/meals designed to replace normal meals during ultra-endurance events such as triathlon, multi-day competition events and long lasting high altitude climbing is recommended. However, the levels should not exceed those of safe daily intake. In this respect it is still an open question whether the RDA established for sedentary people is also adequate for athletes who, due to exercise, may lose substantial amounts of minerals with urine and sweat.

Mineral replacement by adding minerals to rehydration drinks is acceptable as long as the mineral levels do not exceed the upper levels reported for whole body sweat (see Table 3). Although substantial amounts of iron can be lost with sweat there is, to our knowledge, no rationale for replacing this mineral in rehydration drinks during exercise. In general, mineral replacement and/or supplementation in healthy subjects consuming well balanced diets will not enhance performance, but will in certain circumstances contribute to adequate daily intakes. However, in cases of a poor diet composition, or in vegetarian athletes who do not consume meat products known to be rich in specific minerals, an impaired status for some minerals especially iron, zinc and magnesium may develop. A proper education about the role of a well composed diet along with sound supplementation guidelines may then be advised.

Some minerals have been promoted for improvement of performance, because of their specific metabolic influences that may help boost performance. Examples are phosphate and sodium bicarbonate. Because of their use in this way, these substances will be reviewed in more detail in Chapter 10 on nutritional ergogenics.

Key points

- Minerals are important substances for the skeletal structures, as well as for numerous biological actions in the muscles. Impaired mineral status may lead to a reduced bone formation (mineral density) and to muscle weakness.
- Exercise is known to be associated with increased mineral losses, through sweating and also via the urine in the post-exercise phase.

- As with most nutrients, mineral intake depends on the quality of the diet and the amount of energy (food) consumed. High energy consumption leads to increased mineral intake.
- Athletes consuming low energetic diets may be at risk of low mineral intake, especially of magnesium, calcium and zinc.
- Vegetarian athletes may be prone to iron deficiency unless their food choice is appropriate.
- Impaired magnesium status has been suggested to lead to muscle cramp but hard evidence for this is still lacking.
- Mineral supplementation during exercise has not been shown to enhance performance capacity.

7   Trace Elements

The importance of trace minerals (elements) in numerous biological functions, as well as their effect on health and performance, has hardly received any attention until the last two decades. This was mainly due to the lack of analytical methods for measuring and evaluating the role of these elements, which are present in body fluids and tissues in 'micro-quantities'.

Recent technological developments, however, have allowed new insights into trace element status. Aspects of function and bioavailability of some trace elements and the possible effects of exercise on trace element requirements will be discussed briefly in the paragraphs that follow. The trace elements to be discussed are:

- Copper
- Chromium
- Selenium.

TRACE ELEMENT STATUS

In exercise sciences most of the studies in the past have dealt with the macronutrients: fat, protein, CHO and water. The utilization, function and storage of these macronutrients, however, are to a large extent regulated by micronutrients such as specific trace elements and vitamins (5). Shortage of trace elements in the diet may result in impaired trace element status, which is known to influence biochemical and physiological functions and sometimes health. Trace element status is difficult to study. It is possible to obtain samples from serum, tissue, hair, toenails, faeces, urine and sweat. How representative are these samples of whole body status? Analysis of the first four sample types may indicate the status of the pool from which the sample stems. The last three samples may indicate the effect of physical stress on trace element losses and may thus tell us something about the extent of the exercise-induced losses and their possible meaning for daily requirements. However, increased losses do not yield information on the status of different tissues. The growing knowledge over the last 20 years has indicated different sample sites as representative of different trace elements. This make it very complex for any sports doctor to do a simple blood measurement to cover both mineral (see previous chapter) and trace element status.

COPPER

Copper is an essential element for the human body. Copper deficiency has been shown to result in impaired health and malfunctioning. Copper is involved in a large number of enzymes and plays a role in energy metabolism, protein synthesis and protection against free radicals. Additionally, copper influences iron metabolism (131). The activity of erythrocyte superoxide dismutase (SOD), an enzyme eliminating the damaging effect of free radicals, seems to be an objective parameter for copper status (5, 37, 103). Ceruloplasmin, the principal copper binding protein that is present in plasma, may under normal resting conditions give some information on copper status. However, stress is known to release ceruloplasmin from the liver. The latter may thus lead to wrong information about the true copper status in periods of illness or intensive physical exercise leading to exhaustion (5, 37).

Influence of Exercise

Plasma ceruloplasmin as well as serum copper levels have been reported to increase as a result of exercise in some studies, but to remain unchanged or to decrease in other studies (5). Several factors such as differences in training status, type of exercise, degree of plasma volume change or true copper status may account for this. Copper is lost in significant quantities with sweat (93). Therefore, it has been suggested that repeated large sweat losses may impair copper status and that an increased dietary copper intake may be required to offset the losses induced by sustained sweating (4, 5). Thus the normal RDA for copper as determined for sedentary people may be too low for athletes.

Copper Intake

Organ meats, especially liver, are the richest copper sources, followed by seafood, nuts, seeds and potatoes. Milk contains only a low level of copper. The observed copper intake in humans is relatively low, 0.9–1.2 mg/day. Zinc, vitamin C, iron, calcium, protein and dietary fibre as well as a high fructose intake are known to reduce copper absorption and may thus affect copper status (4). The recommended/safe daily intake for copper is given in Table 7 (131).

CHROMIUM

Chromium acts principally in conjunction (as cofactor) with insulin. It is required for a normal insulin activity and consequently a normal regulation of the blood glucose level. Accordingly, experimental chromium deficiency

Table 7 Recommended/safe daily intakes for trace minerals

Source	Copper (mg) f + m	Chromium (μg) f + m	Selenium (μg) f + m
NRC	1.5–3.0	50–200	55–70
DGE	2.0–4.0	50–200	20–100

The data given are derived from NRC and DGE.
NRC = National Research Council, Recommended Dietary Allowances, 1989 (USA).
DGE = Deutsche Gesellschaft für Ernährung, Empfehlungen für die Nährstoffzufuhr, 1991.
f = female, m = male.

results in decreased insulin sensitivity, impaired blood glucose regulation and possibly diabetes. Because of its role in insulin–CHO–energy metabolism, chromium is thought to be of particular importance for people involved in heavy physical work and consuming CHO rich diets. Blood chromium does not appear to be a good marker of chromium status. Urinary chromium losses are a cumulative total of small transitory changes in the blood and appear to be a better indicator of changes in chromium metabolism (3, 5, 97, 131).

Influence of Exercise

Different types of stress, including exercise, infection and physical trauma, are known to exacerbate the signs of marginal chromium deficiency. In the case of exercise this occurs most probably because exercise enhances chromium losses with urine. Additionally, CHO rich diets, especially high glycaemic CHO sources, such as sugars, are known to increase chromium losses with urine. This is most likely an effect of these CHOs on quantitative insulin secretion and subsequent degradation (5). On the other hand it has been shown that CHO loading reduces the rate of trace elements loss, notably chromium and zinc (405). The explanation for this observation is that the level of actual exercise stress influences urinary excretion of these trace elements. In this study, the CHO loading regimen reduced the level of exercise stress as measured by changes in serum cortisol. Losses of potassium, magnesium and calcium were not influenced. These opposite findings on the effects of CHO on chromium levels in the body make it difficult to draw any conclusion. The loss of chromium in sweat has not been quantified using acceptable collection and analytical techniques.

Animal studies have indicated that a poor chromium status is associated with reduced glycogen stores in liver and muscle and that chromium supplementation enhances glycogen storage in this situation. Since endurance performance as well as protein oxidation are influenced by the availability of CHO, it is suggested that sufficient dietary chromium optimizes endurance performance capacity. It has been suggested in various papers that the potentiating action of chromium on insulin is responsible for enhanced incorporation of amino acids in muscle tissue and that this will lead to an increased lean body mass and decreased fat mass (3–5, 35, 37, 96, 103).

Beneficial effects are claimed for chromium picolinate (Cr-pic), which is a fat-soluble compound that easily penetrates cells and enhances insulin internalization (396, 397). Enhanced insulin activity has been observed in adipose tissue samples (395, 396) and glucose uptake was enhanced in yeast cells, due to the presence of Cr-pic. Based on such observations, it has been suggested that Cr-pic maximizes insulin action in skeletal muscle leading to increased muscle mass, reduced fat mass and improved glucose utilization. However, data obtained to support these actions in humans are all of an indirect nature (397, 398) and are criticized for the small number of subjects and experimental design (403). In one study (61) in strength athletes, chromium supplementation in the form of chromium picolinate increased lean body mass and decreased fat mass, as indicated by anthropometric measures only. However, care should be taken with the interpretation of these results, as the chromium status in the test subjects was not controlled and the anthropometric measures used are not a very precise way of determining true muscle mass. Moreover, the patent owners who will benefit from any positive outcome performed the study. In a well controlled study on the effects of Cr-pic supplementation on body composition (400) no effects were observed. Recently the effect of chromium supplementation was reviewed in various papers (400–404). The following conclusions were drawn:

1. Due to enhanced chromium losses and probably marginal intakes, athletes may have an increased chromium requirement. However, chromium deficiencies have not been reported thus far.
2. Data on anabolic action of chromium are presently not convincing.
3. There is no good independent data to support claims made on anabolic effects of Cr-pic.

Independent direct data that show increased protein synthesis as well as insulinic action in human subjects are required to substantiate any beneficial claim made in the current market. Today such data do not exist.

Chromium Intake

The recommended or suggested safe intake of chromium for adults amounts to 50–200 mg/day (131). However, chromium intake in developed Western countries (USA, England, Finland) has generally been found to be lower (5). Chromium absorption varies from 0.3 to 1.0% for inorganic chromium and from 5 to 15% for organically bound chromium such as is present in yeast. Chromium absorption is inversely related to dietary intake at normal chromium intakes but may be decreased by high fibre diets. Chromium absorption is known to influence (inhibit) the absorption of iron and zinc (5, 131).

Important chromium sources are broccoli, oysters, mushrooms, yeast and bran cereals. The chromium content of processed food is known to be increased as a result of processing in metal containers and holding tanks.

SELENIUM

Selenium is an essential component of the enzyme glutathione peroxidase, which regulates the breakdown of hydroperoxides in conjunction with vitamin E. As such, selenium has antioxidant properties. It plays an essential role in scavenging free radicals that are known to appear increasingly in situations of trauma, stress and also during exhausting exercise. Selenium deficits are thought to affect muscle tissue, resulting in cardiomyopathy and muscular discomfort or weakness (37, 97, 103).

Influence of Exercise

Because of its antioxidative function, selenium may help in preventing exercise-induced lipid peroxidation. It may thus help offset the degree of cell damage, most probably in active muscle tissue and in tissues that may undergo a decreased blood flow resulting in local ischaemia. An example of the latter is the gastrointestinal tract, especially the colon (31). No studies have been performed in which the effect of selenium supplementation on oxidative stress has been specifically studied and no data are available on exercise-induced sweat selenium losses (37, 103).

Selenium Intake

The recommended and safe daily intake for selenium is given in Table 7. Seafood, kidney and liver are rich in selenium. Grains and seeds may have a high selenium content, depending on the selenium content of the soil where growth took place (131). Selenium intake in healthy humans normally seems to be adequate. No data are available on selenium intake in athletic populations, which indicates the need for further research.

TRACE ELEMENT REPLACEMENT/SUPPLEMENTATION

Although a well balanced diet containing a variety of fruits, vegetables, grain products, meats and seafood should ensure an adequate trace element intake, it may be concluded from the available literature that healthy people—including athletes—often have a low intake of copper and chromium. These low intakes may cause a poor trace mineral status which may be exacerbated by exercise-induced losses with sweat and urine as well as by low intakes and enhanced losses induced by the high CHO consumption of athletes, especially in endurance events. However, there are no documented reports demonstrating that the overall trace element status of athletes is significantly different from that of normal sedentary people.

Coaches and their athletes want to ensure an adequate daily intake during periods of high intensity training. Because of the possible risks of over-supplementation and the invasive/complex nature of determination of trace element status, athletes and their counsellors should be educated about dietary practices and safe daily intakes. A daily supplementation with a low dose trace element preparation, supplying the recommended daily/safe intake (Table 7), may be advised in periods of intensive training or in any situation where athletes abstain from a normal diet such as during periods of limited food intake combined with intensive training. This may be especially valid for females, vegetarian athletes and athletes participating in low weight class sports (see Table 1). Although a substantial amount of copper may be excreted with sweat, there is to our knowledge no reason to replace these elements with rehydration drinks during exercise. In general, trace element replacement and/or supplementation *will not* enhance performance but may contribute to adequate daily intakes in athletic populations (5, 37, 84).

Key points

- The importance of an appropriate trace element intake and status in athletes has only received a significant interest during the last two decades.
- As is the case for minerals, trace elements are increasingly lost as a result of intensive physical training.
- Trace element losses with sweat (copper) and urine (chromium) may under some circumstances exceed the daily recommended intakes. The composition of the diet may also affect these losses. For example, high CHO intakes, especially of high glycaemic index carbohydrates, have been shown to enhance losses of chromium, whereas diets rich in dietary fibre, often consumed by endurance

athletes and vegetarians, are known to reduce trace element absorption.

- Athletes consuming low energetic diets may be at risk of a low trace element intake.
- Since it is recognized that exhaustive exercise may lead to enhanced tissue/cell damage and regeneration that is associated with an inflammatory process, the importance of selenium, as a component of the free radical defence mechanisms, has received attention. Evidence that supplementation reduces muscle inflammation is still lacking.
- There is no evidence that trace element supplementation induces the development of a larger lean body mass.
- Much research is needed in this field, but it is felt that supplementation with trace element amounts that do not exceed the recommended safe daily intakes will contribute to adequate daily intake in athletes.

8 Vitamins

Vitamins are essential nutrients for the human body. Vitamins are involved in almost every biological function. They serve as coenzymes in many bio reactions, biochemical reactions (including energy metabolism), are involved in protein synthesis and act as antioxidants. The most essential functions of the individual vitamins as well as their role in exercise metabolism and their influence on exercise capacity will be described briefly in the following paragraphs.

VITAMIN STATUS

Several methods are used to determine the vitamin status of the body. Because vitamins function in specific metabolic processes, it is obvious that any significant deficit will affect metabolism and may lead to abnormalities or illness. It is possible to register the occurrence of illness symptoms. However, such symptoms are generally seen as the very last stage of vitamin deficiency. Development of analytical techniques has made it possible to study 'biochemical deficits' that occur at an earlier stage. These measurements include the determination of plasma vitamin levels by high pressure liquid chromatography (HPLC) and by enzymatic stimulation tests. Any such determination of vitamin status is invasive and also expensive. It is important, therefore, that athletes achieve a daily level of vitamin intake that will ensure an optimal vitamin status. This will eliminate the need for invasive tests. Factors that influence vitamin status are food intake and vitamin density of the food, bioavailability (the ability to be absorbed) and losses from the body. The influence of exercise on these factors will be discussed briefly.

INDIVIDUAL VITAMINS AND INFLUENCE OF EXERCISE

In the following paragraphs we will briefly discuss the individual vitamins and reported effects of exercise.

VITAMIN B_1 (THIAMIN)

Vitamin B_1 plays an important role in the oxidative conversion of pyruvate to acetyl CoA, an essential step in the energy production process from CHO.

For this reason the recommended requirements for vitamin B_1 have been related to total energy expenditure and to CHO intake. The RDA is set at 0.5 mg/1000 kcal energy intake (131). It is accepted now that the vitamin B_1 requirement of athletes may be slightly higher due to increased energy and CHO metabolism. Impairment of the maximal oxygen uptake resulting in increased CHO metabolism and lactate production has been shown in humans receiving a vitamin B_1 deficient diet (10). Low intakes of this vitamin as well as biochemical deficiencies have been reported in sedentary subjects and also in athletes, especially in cyclists who consume large amounts of CHO solutions with refined CHO sources (13, 28, 58, 84, 165), although an almost linear relationship with energy intake has been seen (58). There are no controlled studies available on the effect of vitamin B_1 supplementation on performance.

VITAMIN B_2 (RIBOFLAVIN)

Vitamin B_2 is involved in mitochondrial energy metabolism. The National Research Council relates B_2 intake to energy intake. The recommended daily intake is 0.6 mg/1000 kcal, although it is stated that there is no evidence that the requirements increase with increased energy metabolism (131). Few studies have shown that vitamin B_2 requirements for people involved in physical exercise may be increased (28). However, there are no studies that indicate low intakes of this vitamin in athletic populations. Studies in which vitamin B_2 was supplemented in elite swimmers did not show any effects on performance (13, 58).

VITAMIN B_6 (PYRIDOXINE)

Vitamin B_6 plays an important role in protein synthesis. For this reason this vitamin is often assumed to be of crucial importance for strength athletes and bodybuilders. However, there are no data available to support an increased requirement for athletes. Accordingly, studies in which B_6 was supplemented did not improve performance. Some studies indicated performance improvements after supplementation with combined preparations, also including substances that play a role in the citric acid cycle (Krebs cycle; see page 186). However, it is likely that any effect observed was not due to vitamin B_6 but was caused by accompanying substances (13). A dietary B_6 ratio of 0.016 mg/g protein intake appears to ensure acceptable values for B_6 status in adults of both sexes. The RDA is set at 2.0 mg/day for males and 1.6 mg/day for females (131). Recent data are available, indicating insufficient intakes of B_6 in different athletic populations (58).

VITAMIN B_{12} (CYANOCOBALAMIN)

Vitamin B_{12} functions as a coenzyme in nucleic acid metabolism and influence protein synthesis. Endurance cyclists and strength athletes surprisingly often use vitamin B_{12} because it is believed that this compound can have an analgesic effect on muscle soreness when used in mega-doses. Williams (199) and van der Beek (11) reviewed the literature up to 1985 and concluded that there is no evidence for any benefit of supplementation, a conclusion shared by others in more recent reviews. Both oral and parenteral supplementation did not influence any performance related parameters (86, 131). The RDA is 2.0 μg/day (131). Deficits of this vitamin may occur in cases of impaired absorption due to lack of gastric factor (a factor required to make vitamin B_{12} bioavailable), or in subjects who do not consume any meat (only source for B_{12}), as is the case in vegetarians. However, no data are available on vitamin B_{12} intake or on the existence of deficits in athletic populations.

NIACIN

Niacin functions as a coenzyme in NAD (nicotine adenine dinucleotide), which plays a role in glycolysis and is needed for tissue respiration and fat synthesis. The amino acid tryptophan can be converted to niacin: 60 mg of tryptophan has the same response as 1 mg of niacin and is therefore declared as 1 NE (niacin equivalent). Several authors have hypothesized that this vitamin could influence aerobic power, which is an important factor for endurance performance in athletes (199). However, it has been reported that mega-dose intake can also have adverse effects on performance. This may be induced by the inhibiting effect of nicotinic acid on the mobilization of free fatty acid (FFA) from stored triglycerides. During exercise a reduced FFA availability will enhance CHO utilization, which in turn will lead to a higher rate of glycogen depletion. This has been shown to enhance subjective fatigue and to impair performance (13, 85). The RDA has been set at 6.6 NEs per 1000 kcal or at least 13 NEs at caloric intakes of <2000 kcal (131). No data are available on niacin intake or on deficiencies in athletic populations, or on effects of niacin supplementation on performance.

PANTOTHENIC ACID (PA)

Pantothenic acid is a component of acetyl CoA, the intermediate citric acid cycle metabolite of CHO and fat metabolism. Williams stated in 1985 that some reports suggested a beneficial effect of PA supplementation but that conclusive data were not available (199). This has not changed until now. No data are available on PA intake or on deficiencies in athletes.

Supplementation with pharmacological doses as high as 1 g/day did not result in any performance improvement (13). The National Research Council concludes that there is insufficient evidence to set a RDA for pantothenic acid. The safe daily intake level is assumed to be 4–7 mg (131).

FOLATE

Folate functions as a coenzyme in amino acid metabolism and nucleic acid synthesis. The RDA for folate amounts to approximately 3 μg/kg body weight, resulting in a daily RDA of 200 μg for males and 180 μg for females (131). There are no controlled studies available on the effect of folate supplementation on physical performance, or on folate intake in athletes (13). Plasma folate levels, which may reflect folate intakes, were observed to increase in Tour de France participants, who ingested substantial amounts of vitamin preparations (164). Williams (202) cited recent research that folate supplementation would restore normal folate status to runners who were folate deficient, but did not improve performance capacity.

BIOTIN

Biotin is an essential part of enzymes that transport carboxyl units and fix carbon dioxide in tissues. The conversion of biotin to active coenzyme depends on the availability of magnesium and ATP. Biotin plays an essential role in CHO, fat, propionate and branched chain amino acid metabolism. Biotin is produced in the lower intestine by microorganisms and fungi. No data are available on the quantitative absorption and there are insufficient data to establish a RDA for biotin. A range of 30–100 μg/day is provisionally recommended as a safe daily intake for adults (131). There are no studies available on supplementation effects, or on biotin intake or on intestinal synthesis in athletes (13).

VITAMIN C (ASCORBIC ACID)

Vitamin C is probably the most studied vitamin. Vitamin C is a water soluble antioxidant. It scavenges free radicals that cause cell damage and protects vitamin E, another antioxidant, from destruction. It participates in many enzymatic reactions by acting as an electron transmitter, and is involved in the synthesis of collagen and carnitine (the latter is needed for the transport of long chain fatty acids across the mitochondrial membrane prior to oxidation). Vitamin C enhances iron absorption in the gut. It is also needed for the biosynthesis of some hormones (14, 68, 131). Early studies performed during the Second World War showed that insufficient vitamin C lowered physical performance capacity in soldiers and increased the sensation of exhaustion and muscle pains during and after hard physical work. However, many of the studies performed at that time have now been criticized for their poor methodology, control and statistical design. More recently well controlled double blind studies have shown that a state of moderate vitamin C deficiency does not reduce physical performance in single intensive bouts of exercise. There are some indications that vitamin C may enhance the rate of heat acclimation (13). This may be of benefit to athletes involved in endurance competitions in the heat around different parts of the world. However, vitamin C supplementation did not improve performance in controlled studies. A study in long distance runners has shown that the supplementation of vitamin C prior to the run results in a decreased occurrence of respiratory infections (236). In general, vitamin C intake in athletes is sufficient, with the exception of individuals consuming a low caloric diet (12, 13, 28, 58, 68). Vitamin C is also a powerful antioxidant. This aspect is dealt with in Chapter 9.

VITAMIN E (ALPHA-TOCOPHEROL)

Vitamin E is an antioxidant and scavenges free radicals to protect cell membranes from lipid peroxidation. It functions in concert with vitamin C, beta-carotene and selenium, and also protects red blood cells from haemolysis (14, 131, 171). In the period 1970–1980, special attention was given to this vitamin after reported beneficial effects of its supplementation on oxygen consumption and physical performance. As is the case with vitamin C, many of these studies were also not well controlled or suffered from poor statistical design. Critical analysis of the literature and more recent results from well designed double blind studies did not bring any solid evidence for performance improvement (13, 14, 84, 171, 206). It has been observed that endurance athletes in general have low vitamin E serum levels. This may be an indication of either marginal vitamin E supply with food or increased usage in antioxidant defence mechanisms.

There are indications that vitamin E supplementation elevates the testosterone/cortisol ratio suggesting that vitamin E has a stress reducing effect on the body. It is able to reduce lipid peroxidation in both animals and humans as measured by an enhanced appearance of penthane in exhaled air. Studies at high altitude indicate that vitamin E can influence metabolic performance parameters and reduce penthane (an indirect marker of free radical induced cell damage) exhalation, suggesting that vitamin E may have a protective effect. However, it is not known which tissues undergo lipid peroxidation most during exercise. It may be that the most important site is tissue that is prone to some ischaemia during exercise, such as the gastrointestinal system, but not muscle.

Since it became possible to measure the effects on free radical pathology, attention has been given to the antioxidant properties of vitamin E. This aspect is dealt with in Chapter 9.

VITAMINS A, D AND K

Although the importance of the fat soluble vitamins A, D and K for health is beyond doubt (131), there are no studies available which indicate any significant effect of these vitamins on biochemical or physiological parameters concerned with physical performance capacity. Since these vitamins are potentially toxic, when taken in high doses for a prolonged period of time (with the exception of vitamin K), and daily intake in Western civilized countries is generally sufficient, there is no need for supplementation (13, 83, 84, 199).

Vitamin K serves a function in bone mineralization. This was found after the observation that the intake of anticoagulant drugs (vitamin K antagonists) influences bone formation processes. Accordingly, the role of vitamin K on bone formation and the prevention of osteoporosis is currently under study (237). Recently the effect of vitamin K supplementation, 10 mg/day, has been studied in eight female endurance athletes, four of whom had been amenorrhoeic for more than one year, while the remaining four had been using oral contraceptives. Such female endurance athletes have depressed oestrogen levels and may develop mineral loss from bone to an extent comparable to postmenopausal women. It was observed that in all subjects increased vitamin K intake was associated with a 15–20% increase in markers of bone formation and a parallel decrease of 20–25% in markers of bone resumption, suggesting an improved balance between bone formation and loss (238). Further research seems to be justified to determine whether long-term vitamin K supplementation is of benefit to bone health for the female athletic population.

Vitamin Intake

Some aspects of intake of the individual vitamins in athletes have been discussed in the previous paragraphs. Here we will discuss some general influences on daily vitamin intake. Vitamins are present in a wide variety of fresh unprocessed foods such as vegetables, fruits, berries, tubers and grains. A normal well balanced diet composed of a variety of foods is therefore believed to supply all necessary vitamins in sufficient quantities. However, in some situations the intake may be lower than the current RDA. Such a situation may occur when low energetic diets or unbalanced diets are consumed. This first situation occurs frequently in athletes who compete in low weight categories and follow weight reducing diet-training programmes. Alternatively, athletes who have to maintain a low body weight for prolonged periods of time such as female dancers and gymnasts may be prone to low vitamin intakes (Table 1).

Figure 38 Vitamin E supplementation may improve oxygen uptake and performance at high altitude, but more studies are required before a supplementation recommendation can be justified (photo ARPE)

Figure 39 Mountain running leads to mechanical microdamage in muscle fibres. It is suggested that recovery from muscle cell damage is improved when antioxidant vitamins are supplemented.

Figure 40 French fries are popular all around the world. Unfortunately many young athletes consume snack foods regularly between meals. This may affect the supply of essential nutrients in a negative way

VITAMIN RESTORATION AND SUPPLEMENTATION

As discussed in the chapters on minerals and trace elements (Chapters 6 and 7), individuals at potential risk of marginal micronutrient supply are those who consume low caloric diets for prolonged periods of time. A relatively low supply of vitamins may also occur when large amounts of processed foods constitute the major part of the daily diet. This has been observed to be the case in endurance athletes who ingest relatively large amounts of refined CHO as energy drinks during their sports events (23, 58, 165). The reason for this has been discussed in Chapter 2. In both situations the required micronutrient density (i.e. the amount of vitamins present per 1000 kcal energy intake) is higher than can be achieved in the diet. In these situations athletes may be advised to take a daily vitamin–mineral–trace element supplement (not more than 1–2 times RDA/daily) to enhance micronutrient density and secure an appropriate intake.

In industrially processed products/meals, vitamins are often added to replace processing-induced losses (restoration) or to increase the vitamin content slightly above normal (enrichment/fortification). In general, vitamin restoration or fortification of energy dense processed foods like energy beverages or bars, as well as supplementation with pills or capsules, will not enhance performance (13, 195) but may contribute to an adequate daily intake.

Daily intake of a low dose vitamin supplement, or a nutrient preparation that supplies not more than the recommended daily intake (Table 8), in addition to the normal diet, is recommended in periods of intensive training

Table 8 Recommended dietary allowances for vitamins

Vitamin	Age: males			Age: females		
	NRC 15–18 DGE 15–18	19–24 19–25	25–50 25–51	15–18 15–18	19–24 19–25	25–50 25–51
Vit B$_1$ (mg)	1.5/1.6	1.5/1.4	1.5/1.3	1.1/1.3	1.1/1.2	1.1/1.1
Vit B$_2$ (mg)	1.8/1.8	1.7/1.7	1.7/1.7	1.3/1.7	1.3/1.5	1.3/1.5
Niacin (mg)	20/20	19/18	19/18	15/16	15/15	15/15
Vit B$_6$ (mg)	2.0/2.1	2.0/1.8	2.0/1.8	1.5/1.8	1.6/1.8	1.6/1.6
Folate (μg)	200/300	200/300	200/300	180/150	180/150	180/150
Vit B$_{12}$ (μg)	2/3	2/3	2/3	2/3	2/3	2/3
Vit C (mg)	60/75	60/75	60/75	60/75	60/75	60/75
Vit A (μg) RE	1000/1000	1000/1000	1000/1000	800/900	800/800	800/800
Vit D (μg)	10/5	10/5	5/5	10/5	10/5	5/5
Vit E (mg) TE	10/12	10/12	10/12	8/12	8/12	8/12
Vit K (μg)	65/70	70/70	80/80	55/60	60/60	65/65
Pantothenic acid	^{a}/8	^{a}/8	^{a}/8	^{a}/8	^{a}/8	^{a}/8

The data shown are derived from NRC/DGE.
NRC = National Research Council, Recommended Dietary Allowances, 1989 (USA).
DGE = Deutsche Gesellschaft für Ernährung, Empfehlungen für die Nährstoffzufuhr, 1991.

or in any situation where athletes abstain from a normal diet such as during periods of limited food intake combined with intensive training (especially in females, in vegetarian athletes, and in weight class sports participants, see Table 1). Although the use of mega-doses of vitamins by athletes is often defended with the argument that substantial amounts of vitamins may be lost with sweat and urine, there are no scientific data supporting this. Sweat vitamin losses are in general negligible (13, 28, 84). Therefore, the use of high vitamin doses should be discouraged because of potential undesired side effects (3) and possible negative interactions with other micronutrients (109). Performance benefits resulting from vitamin-mineral supplementation are unlikely to occur (479, 480)

VITAMIN AND MINERAL USE BY ATHLETES

The use of vitamins and minerals as ergogenic substances has been reviewed recently by Sobal and Marquart (355). They analysed 51 studies that provided data on the quantitative use of such preparations in over 10 000 male and female athletes. The mean prevalence of use was 46%. Elite athletes used more supplements than lower ranked athletes, and women used more than men. Of the 51 studies, 32 provided information about the type of supplement used. Multivitamins and iron were most frequently taken, followed by B vitamins, vitamin E, calcium and vitamin A. Women athletes particularly took iron, sometimes in high amounts. In most cases, iron supplements were used on a regular daily basis, rather than occasionally or weekly.

Some studies found that some elite athletes exhibit extreme levels of supplement use. Some Olympic athletes were found to consume as many as 14 different types of supplements/day and 63 pills/day (355). Faber and Spinnlerbenade (352) described an average supplement pill intake of 19 per day, with one athlete consuming as many as 87 pills per day.

Reasons for supplement use were various: performance enhancement; providing extra energy; illness prevention; vitality enhancement; insurance to get a daily adequate intake of important vitamins and minerals; support of training adaptations; and muscle development.

Information on why specific supplements are used is most frequently obtained through coaches although parents, doctors and peers also are important. A survey by Parr et al. (357) revealed that the opinion among athletes, coaches and trainers in the US is that the trainer is expected to have the primary responsibility for the athlete's nutrition.

Media and industrial advertising have great influence on decision making (353, 354). The myth that vitamins will give extra energy, as is often claimed in advertising, is misleading, but is accepted by part of the athletic population who are persuaded to take extra vitamins when muscle weakness or fatigue is present (351).

The myth that L-carnitine enhances fat oxidation is plausible enough to persuade many endurance athletes as well as people willing to slim and reduce their body fat to use this supplement (368). A number of studies have analysed the nutritional knowledge of athletes and their trainers (350, 351, 353, 354, 356, 357, 359–361). These studies tend to show that a better level of information about aspects of daily nutrition and about efficiency of food supplements leads to a more reasonable ingestion of supplements.

Basic knowledge about the type of sport in which the athlete is involved and the consequences that this has for food consumption is considered to be important. Many endurance athletes, for example, do not realize that they ingest substantially more food, on a daily basis, than non-endurance athletes. The consequences of this larger food consumption are also an increased consumption of most micronutrients and of proteins. In this respect we had observed the opinion among coaches involved in professional cycling that the cyclists need extra protein to be able to complete the Tour de France. However, although this is factually right, it was observed that at the high level of food intake in these athletes and a constant level of approximately 12 en% protein intake, the daily protein intake level of 2.4 g/kg bodyweight was in excess of the enhanced requirements (165).

These examples show that proper, science based information is essential for the understanding of the interrelation between sport events and nutritional needs, especially at the level of decision making/of opinion makers such as coaches, trainers and parents.

Key points

- Vitamins are essential cofactors in many enzymatic reactions involved in energy production and in protein metabolism.
- Any shortage of a vitamin may lead to suboptimal metabolism, which in the long term may result in decreased performance or even illness.
- Some vitamins act as antioxidants and there is accumulating evidence that nutritional antioxidants may help optimize the protective role for the maintenance of tissue/cell integrity.
- Vitamin supplementation has been shown to restore performance capacity in cases of a vitamin deficit and to reduce tissue damage due to free radical production.
- Vitamin supplementation with quantities exceeding those needed for optimal/blood levels has not been shown to improve performance.
- As is the case for minerals and trace elements, athletes involved in intensive training, but consuming low energetic diets, are most prone to marginal vitamin intakes.

- It can be concluded that vitamin restoration of energy dense processed foods or supplementation with preparations will not enhance performance but may, in athletic populations, contribute to adequate daily intakes.
- Daily intake of a low dose vitamin or combined vitamin–mineral–trace element preparation, supplying the recommended safe daily intake, may be advised in periods of intensive training or in any situation where athletes have to abstain from a normal diet.

9 Antioxidants and Exercise Induced Free Radicals

It is generally recognized that free radicals are formed during oxidative energy production processes. During these processes oxygen becomes reduced to form water, while adenosine triphosphate (ATP) is formed from adenosine diphosphate (ADP).

Certain physical restrictions dictate that oxygen can only receive one electron at a time while four electrons are required to produce water. This univalent pathway of oxygen reduction transiently leads to the production of free radicals. Addition of one, two or three electrons to molecular oxygen leads to the production of superoxide (O_2^-) hydrogen peroxide (H_2O_2) and hydroxyl radical (OH$^{\cdot}$) respectively. It has been estimated that about 2 to 5% of the total electron flux during normal metabolism 'leaks off' to generate free radicals. Accordingly, intensified respiration during sports activity is accompanied by an enhanced free radical production that can be further augmented by increased body temperature and increased stress hormone levels.

Fortunately the human body has a number of ways to eliminate free radicals once they are formed rapidly. There are a number of enzyme systems that are able to quench most of the radicals. Additionally, there are a number of antioxidant compounds that circulate with blood and/or are present in tissues and cells that can help reduce free radical damage. Among such compounds are the vitamins beta-carotene, C, E as well as hypoxanthine resulting from the degradation of adenosine monophosphate during intense exercise.

Interestingly, trained individuals have increased levels of these enzymes, which points to two important aspects: (i) exercise leads to higher levels on free radicals; (ii) the body responds to this with a physiological adaptation being an upregulation of both number and activity of enzymes in defence systems.

In this respect it is relevant to question whether the defence systems of the body are appropriate under all circumstances. For example, is an intense performance while being insufficiently trained and having a non-adapted enzymatic system of potential damage to health? Or, can the absolute top performance that an elite athlete delivers, for example in a marathon, be improved by ingesting oral antioxidants with the goal to reduce cell damage and maintain optimal muscle cell contractile capacity? Again, can the daily training quality and volume be improved by the ingestion of oral antioxidants? The latter question is relevant as it has been observed that the muscle pain that can occur after intense exercise (known as delayed onset of muscle soreness (DOMS)), is related to an inflammation process. In this process macrophages help to break down the damaged muscle fibres in order to initiate a repair process in which an inflammatory reaction is involved. The increased levels of free radicals parallel the painful process which is known to impair muscle strength. This observation has led to much speculation about the possible benefits of antioxidant supplementation on muscle function, muscle recovery and the quality of life of the athlete.

The short review that follows here deals with basic aspects of free radical production, the effects on antioxidants and the effects of exercise. As is the case with other chapters, the documentation presented here is based on existing reviews and original research papers which deal with this topic. Most of the introductory part of this chapter is based on citations from the following excellent papers: Conning-British Nutrition Foundation (280); Muggli (278), National Research Council (279). The interested reader can find more detailed information in other reviews that include specific aspects of exercise (14, 271–275). The most comprehensive recent reviews are those of Sen *et al.* (276) and Li Li Ji (277).

WHAT IS A FREE RADICAL?

Atoms consist of a nucleus with electrons in 'orbit' around the nucleus. The number of electrons is the same as the number of protons in the nucleus but the electrons must be arranged in layers (or 'shells') so that the inner layer can contain no more than two, the next no more than eight, the third no more than 18. Atoms are most stable when the electrons in each shell are paired, and the electrons of each pair orbit in opposite directions. The chemical reactions required to maintain the function of living processes of land-based animals usually take place at atmospheric pressure and at body temperature. The same reactions, if conducted in a laboratory, would require raised temperatures and pressures. Living organisms can achieve these chemical syntheses and breakdowns by enzyme systems that change the electron distribution of the molecules involved. One type of product of such an effect is the 'free radical'.

A free radical is an atom or molecule, capable of existing independently for an extremely short period of time, which contains one or more unpaired electrons. An overview of different free radicals also called reactive oxygen species is given in Table 9. Some free radicals are very reactive chemically and thereby carry the potential for extensive damage to the organism in which they are generated. Free radical reactions involve the donation or acquisition of a single electron. This tends to create another radical

Table 9 Reactive oxygen species

Radical	Name	Typical biological target
$O_2{}^{\cdot-}$	Superoxide	Enzymes
H_2O_2	Hydrogen peroxide	Unsaturated fatty acids
$HO^{\cdot}$	Hydroxyl	All biomolecules
$R^{\cdot}$	R-yl	Oxygen
$RO^{\cdot}$	R-oxyl	Unsaturated fatty acids
$ROO^{\cdot}$	R-dioxyl (R-peroxyl)	Unsaturated fatty acids
$ROOH$	Hydroperoxide	Unsaturated fatty acids
$^1O_2$	Singlet molecular oxygen	H_2O
$NO^{\cdot}$	Nitroxyl	Several

whereupon the process may be repeated. It is the nature of free radical reactions that they may create *chain reactions* and it is the function of the defensive system to stop such propagation.

FREE RADICAL PRODUCTION

In our body various processes and reactions can produce free radicals. Table 10 divides these into two main sections: (1) those formed as a result of the impact of radiation; (2) those formed by reduction–oxidation (redox) reactions involving the transfer of an electron.

Oxidizing free radicals may initiate or extend cell injury by removing a hydrogen atom from, for example, a polyunsaturated fatty acid in a biomembrane, initiating the degradative process of lipid peroxidation.

They may also add across unsaturated centres in molecules to give covalently bound adducts that may have a strongly disturbed biological function. As such, free radicals can affect protein and nucleic acid

Table 10 Major mechanisms resulting in the formation of reactive free radical intermediates

Formation
1. By the impact or absorption of radiation, or both:
    (a) high energy or ionizing radiation
    (b) ultraviolet radiation
    (c) thermal degradation of organic material
2. By electron transfer ('redox') reactions:
    (a) catalysed by transition metal ions
    (b) catalysed by enzymes

metabolism, biomembrane integrity, enzymes and thus tissue function and pathology.

Meanwhile a large number of diseases and toxic cell injuries are known to be associated with free radical production. Unclear, however, in many of these is whether the free radicals cause these pathologies or result from an initiated pathological process and further worsen this process.

Results from numerous studies of the last decade show that the generation of oxygen free radicals is an essential feature of normal oxidative metabolism in which a large number of enzymes are involved. However, this generation is increased in situations of oxidative metabolic stress such as ischaemia, especially following trauma but most probably also during highly intensive exercise and gut ischaemia or muscle tissue damaging activities.

In this respect scientific interest over the years was focused on the effect of antioxidant status on free radical damage. *In vitro* studies had shown that lack of antioxidants in a biological system resulted in significant free radical pathology, whereas addition of antioxidants to the system reduced pathology partly or totally. Epidemiologic studies indicated that low dietary intake of antioxidant nutrients is associated with increased incidence of tissue pathology, in particular lung (beta-carotene), breast, colon and prostate (selenium) or overall cancer (vitamin C). The organism must be equipped, therefore, with potent defence mechanisms to prevent such damage. The human body has a variety of mechanisms to protect itself against the effects of free radicals—enzymes, scavengers and antioxidants.

Enzymatic defence mechanisms are mainly within the cell and have minerals or trace elements as cofactors. Non-enzymatic defence mechanisms can be found as specific proteins in blood, which bind minerals or trace elements with a high free radical reaction initiating potential. Lastly, the body benefits from small non-enzymatic molecules which are water or fat soluble and which circulate throughout the body with blood and are also present in all tissues. Antioxidant nutrients belong to this class. Table 11 gives an overview of the major antioxidant defence mechanisms in biological systems.

Various publications give more details about specific antioxidants and cofactors: beta carotene (281, 282, 284, 295, 298), vitamin C (287), phenols (300), vitamin E (293, 294, 296, 297), Q10 (6, 302), copper (291), selenium (283, 292), hypotaurine (286).

ENZYMATIC SYSTEMS

Superoxide dismutase (SOD) catalyses the conversion of superoxide radical to hydrogen peroxide and oxygen. It therefore works in conjunction with

Table 11 Major antioxidant defences in biological system

A. Enzymatic (mainly intracellular)	
SOD*	Cu-Zen enzyme, Mn enzyme
Catalase	Haem enzyme
GSH peroxidase	Selenoenzyme, non-Se-enzyme
GSH transferase	Detoxifies carbon-centred electophilic agents
GSH reductase	Regenerates GSH
B. Non-enzymatic proteinaceous (blood plasma)	
Ceruloplasmin	Binds Cu, ferro-oxidase activity, $O_2^{\cdot-}$ scavenger
Transferrin-	Binds Fe
Albumin	Binds Cu
Haptoglobin	Binds free haemoglobin
Haemopexin	Binds free haeme
C. Non-enzymatic small molecules	
1. Water soluble	
Ascorbic acid (vitamin C):	Most important water soluble antioxidant at normal oxygen pressure. Scavenges $O_2^{\cdot-}$, OH, H_2O_2 and $^1O_2$
Glucose	
Uric acid	
Bilirubin	
GSH (Glutathione)	
2. Lipid-soluble	
α-Tocopherol (vitamin E):	Most important lipid soluble antioxidant in blood. Chain-breaking antioxidant. Scavenges $O_2^{\cdot-}$, OH, H_2O_2 and $^1O_2$
Ubiquinol-10	
β-Carotene:	Most efficient $^1O_2$ quencher. Antioxidant at low oxygen pressure
Lycopene	
Lutein	
Zeaxanthin	

*Super Oxide Dismutase

catalase and glutathione peroxidase. The SOD in mitochondria is manganese dependent. The enzymes present in the cytoplasm are zinc and copper dependent.

Catalase is the enzyme that deals with hydrogen peroxide in specialized compartments called peroxisomes. Catalase, in effect, promotes the transfer of electrons from iron to form water and oxygen. Catalase is iron dependent.

Thiols are compounds that contain sulphydryl groups (—SH). Many proteins contain such groups but the most abundant non-protein thiol is glutathione (GSH). The bulk of this is in the cytoplasm of the cell, but about 15% is in the mitochondria where it is highly concentrated. GSH may react directly with a radical but its ability to do this is much increased by another enzyme, glutathione-S-transferase (GST).

Glutathione is also involved with the destruction of hydrogen peroxide in the cell. A selenium dependent enzyme, glutathione peroxidase, transfers hydrogen to hydrogen peroxide to form water. The oxidized glutathione, in the form of a double molecule or dimer (GSSG) is reduced by the enzyme GSH reductase to restore GSH.

ANTIOXIDANT COFACTORS

Several components of the enzymatic defence systems require certain minerals and trace elements as an integral part of their structure to function properly. Catalase is iron dependent, SOD can be zinc, copper or manganese dependent and glutathione is selenium dependent. It is expected that a poor status of these food substances may impair the enzymatic defence systems.

NON-ENZYMATIC PROTEINACEOUS SYSTEMS

The body possesses transport proteins, which bind minerals and trace elements after absorption or liberation into blood. Since these substances could be harmful in the free form, these proteins can be seen as a defence mechanism. With low supply of the minerals the binding activity of these proteins increases whereas when being saturated the binding activity decreases. Problems occur whenever either the body is overloaded with a specific mineral, e.g. iron, so that the binding capacity is exceeded and free iron levels increase, or when, due to malnutrition of especially protein and energy the quantity of the binding protein decreases.

NON-ENZYMATIC SMALL MOLECULES: ANTIOXIDANTS

To this group of defence mechanisms belong all antioxidant vitamins and other substances which decrease oxidation in biological systems such as ubiquinone, lutein, lycopene, flavonoids, taurine and plant phenols and indoles. The important issue of these antioxidants is that increasing the dietary intake can raise their concentration in blood and tissues. Concomitantly the antioxidant potential of the body will increase or decrease depending on the supply with daily food.

ANTIOXIDANTS

Antioxidants are compounds that readily donate electrons or hydrogen without themselves being converted into highly reactive radicals. There are several classes of nutritional compounds that can do this:

- *Vitamin E* is a fat soluble vitamin (alpha-tocopherol) present in cell membranes and lipoprotein particles. It readily reacts with hydroxyl radicals, donates its own hydrogen and thereby terminates the chain reaction, which could be so damaging to membranes. There is evidence that the vitamin E, which has been used for this purpose, can be 'restored' by the action of ascorbic acid (vitamin C). Other forms of vitamin E (i.e. other tocopherols) are less active as antioxidants.
- *Vitamin C* is a water soluble vitamin (L-ascorbic acid) known to be involved in a number of important biological syntheses and in the absorption of iron and copper. Vitamin C is used worldwide as an antioxidant in food processing. There is some evidence that, in addition to the roles mentioned, it is required for the generation of nitric oxide (no) by macrophages. No radicals may be involved in the elimination of bacteria. There is evidence that vitamin C 'protects' vitamin E from destruction in foods after the change of vitamin E into a radical, $E^{.}$. Thus protective interactions of antioxidants may exist.
- *Beta-carotene or pro-vitamin A* is an antioxidant and fulfils this role without conversion to vitamin A. Beta-carotene is particularly effective against singlet oxygen, a very reactive species in which an electron has been 'excited' to an orbital above that which it normally occupies. Other carotenes such as lycopene and lutein are under investigation for similar radical trapping activity. Total antioxidant capacity of blood plasma is made up of a variety of factors of which vitamin E and vitamin C play a small role and uric acid and protein thiols play a major role. From this observation one may conclude that no single antioxidant has prime importance for optimal health but there is a synergistic action of antioxidants, cofactors and enzymatic defence. Additionally, substances with antioxidant capacity derived from plant food may be of prime importance in the whole system.
- *Plant phenols and indoles.* Plants contain a large number of phenolic compounds that may act as electron donors. Few of these have been studied extensively for their role as dietary antioxidants. Some indoles are known to inhibit the activity of those cancer-causing chemicals that must be converted into electrophiles before the cancerous effect becomes evident. It is not clear whether the mechanism is a direct effect or involves the induction of the enzymes involved in the defence systems.

- *Organosulphur compounds.* Allyl di- and trisulphides, chemicals found in onions and garlic, are known to induce the synthesis of glutathione-S-transferase (GST). They act, therefore, not primarily as antioxidants but as a stimulus to antioxidant mechanisms.

FREE RADICALS IN SPORT

Highly intensive sport performance is characterized by a number of events, which make increased free radical production and related cell damage most probable. Oxygen consumption for aerobic energy production increases about 20-fold and so does free radical production since both processes are quantitatively interrelated.

Additionally, free radicals may result from energy depletion in skeletal muscle during which ATP is broken down to ADP→AMP→ hypoxanthine, which finally leads to the formation of xanthine and uric acid in red blood cells and endothelial cells, resulting in the liberation of free radicals. This is the xanthine oxidase (XO) pathway. Also the auto-oxidation of catecholamines as well as the production of nitric oxide, substances that are increased during exercise, lead to free radical production.

From animal experiments as well as from surgery in humans it is known that a restriction of blood flow, followed by reperfusion (restoration of blood flow) is accompanied by an enhanced production of free radicals. Similarly, it may be that a significant reduction of the blood flow to the gastrointestinal tract, as takes place during endurance events in a dehydrated state, may cause similar effects. In marathon runners, a condition of gut ischaemia and gut mucosa necroses, leading to bloody diarrhoea after the race, has been observed and one may speculate that free radicals are associated with the damage of the epithelial gut cells. Also, iron released from red blood cells during haemolysis may induce free radical formation (290).

Muscle soreness after an intensive bout of exercise in less well trained subjects may be linked to free radicals. The micro-trauma (disruption at the Z band level of the sarcomeres) which results from acute overload cannot be avoided by antioxidant systems, because it is mechanical in nature. The repair process of mechanically damaged muscle fibres, however, involves an inflammatory process, which causes muscle pain, stiffness and loss of muscle strength, especially in the period 2–5 days after the sport event. It is suggested that free radicals play an important role during this inflammatory process and that supply with adequate amounts of antioxidants may lessen both the severity and the duration of this delayed muscle soreness (288).

Endurance exercise in polluted air, such as running a major city marathon on a hot summer day in the smog, has been suggested to lead to damage to the lung tissue induced by ozone. Free radicals are suspected to be the mediating mechanism. Accordingly, vitamin E supplementation is suggested to reduce such damage and lung function impairment (285).

Training is known to significantly increase the activity of the enzymatic defence mechanisms, a normal physiological adaptation (289, 299). One may speculate that such an adaptation in itself may be sufficient to offset possible effects due to increased free radical formation. If not, regular intensive exercise would lead to impaired health and body function.

The conclusions and consensus as listed below can be obtained from the above-cited reviews, in particular those of Sen et al. (276) and Li Li Ji (277).

Key points

- Free radicals are involved in the aetiology of cell damage and tissue pathology.
- The body possesses several defence mechanisms against free radicals, enzymatic and non-enzymatic, including nutrient derived cofactors
- Free radical production during and after exercise is increased as a result of oxidative and metabolic stress, micro-trauma and ischaemia.
- The body's defence mechanism capacity depends on nutritional status, especially the adequate supply of substances with free radical scavenging properties. Free radical damage to cells and tissues is assumed to be aggravated in cases of inappropriate defence mechanisms.
- Marginal supply with cofactors, such as selenium, zinc, copper and manganese may limit enzymatic adaptations. Hard data, from athletic populations, showing that intake of such cofactors is insufficient and that this may impair free radical defence mechanisms is lacking, however.
- Antioxidant supplementation may have an effect in cases of impaired defence mechanisms as seen with marginal nutritional intake leading to antioxidant vitamin or cofactor depletion.
- In well trained healthy athletes there is no evidence of impaired body defence mechanisms
- Vitamin E supplementation has been shown to reduce enzyme markers of tissue damage in the post-exercise phase. The effect of this observation on athletic performance or health status of the athlete remains unclear. Performance remains unaffected.

- Vitamin C and Q10 when given in higher dosages can work as a pro-oxidant, which will potentiate free radical production. The role of both compounds on performance enhancement has not been proven.
- There is lack of evidence for the role of beta carotene in performance and in antioxidant defence mechanisms in the exercising athlete.
- Glutathione supplementation has been shown to improve performance in animal studies and to prevent exercise induced oxidation of GSH. As such GSH supplementation may be promising for further research to define possible benefits for the athlete.

IV Nutritional Ergogenics and Metabolism

10   Nutritional Ergogenics

Nutritional ergogenics are food substances that have a performance enhancing effect. This can be physical as well as mental. Since the control of anabolic steroids and other illegal performance enhancing substances has been intensified, a variety of nutrients have been put forward as effective, safe and legal alternatives. In this chapter some potential nutritional alternatives to illegal drugs, as well as some food supplements marketed for athletes, will be discussed.

RIBOSE

Repeated bouts of intense all-out exercise result in a significant production of lactate and ammonia while rapidly depleting muscle glycogen and phosphocreatine levels. Phosphocreatine breakdown and glycolysis are the primary pathways of ATP production during short strenuous exercise. At very high exercise intensities the extensive rate of ATP breakdown may exceed the ability of the creatine kinase reaction and glycolysis to rephosphorylate ADP. This results in an increase of the intracellular ADP content. A fraction of this ADP is eventually degraded to IMP, which may be further converted to inosine and hypoxanthine. These compounds may be washed out from muscle and accordingly lead to a reduction of the total muscle adenine nucleotide pool (TAN) (241, 242, 249). Following exercise TAN is partially replenished via the purine salvage pathway, which involves the successive reverse conversion of hypoxanthine to inosine monophosphate (IMP) and further to adenosine monophosphate (AMP). The remaining fraction needs to be recovered by *de novo* nucleotide synthesis. However, compared with the purine salvage pathway the latter process is slow. Hence muscle ATP content can be decreased for several days after high intensity exercise (241, 249, 250). It has been hypothesized that this inefficient recovery is caused by a low activity of the rate-limiting enzyme (glucose-6-phosphate dehydrogenase) in the pathway of adenine nucleotide synthesis. This enzyme is critical for the production of phosphoribosyl-pyrophosphate, a compound that is a precursor for IMP production and is required for both the *de novo* nucleotide synthesis and the recovery of AMP from its degradation products in the salvage pathway. The activity of this enzyme is difficult to upregulate, even in a situation of depressed TAN levels. There is some evidence to suggest that this limitation can be overcome by supplying ribose to the muscle. Animal studies have indicated that *supraphysiological concentrations* of ribose enhanced ATP synthesis. Furthermore, increasing ribose provision by intravenous ribose infusion was found to markedly enhance the recovery of myocardial ATP content as well as the functional capacity in various animal models of myocardial ischaemia (243, 245, 247, 248, 251–254).

Oral intake ribose has been shown to be rapidly absorbed from the intestinal tract, and is well tolerated even at very high dosages (>100 g per day) (240) and during exercise (239). Following absorption, ribose is rapidly and extensively metabolized, the principal fate being conversion in the liver to glucose via the pentose phosphate pathway (246, 258). Furthermore, ribose can also be transported into muscle cells to feed the nucleotide synthesis pathway. It has been observed that ribose supply to skeletal muscle of rats in an experimental contraction model improved the rate of adenine nucleotide recovery from degradation products four- to six-fold (250). Early animal studies have shown that the recovery of TAN after a period of ischaemia in the heart as well as liver is significantly enhanced by supply of ribose (256). This observation has led to studies in cardiac patients who suffer from heart ischaemia. It was shown in these studies that the heart's tolerance to ischaemia improved and that symptom free treadmill walking time increased by > 30% (257). However, in most human studies on ribose supplementation neither measurement of blood ribose levels nor of changes in TAN was done. As such, direct evidence in humans is entirely missing.

The first and currently only available double blind randomized human performance that also measured the changes of TAN in muscle as well as the ribose level in blood did not find any effect on performance (255). In this study muscle power output was measured during dynamic knee extensions with the right leg on an isokinetic dynamometer before (pre-test) and after (post-test) a 6 day training period in conjunction with ribose (R, 4 × 4 g per day taken within a period of 4 h 'around' the exercise bouts, n = 10) or placebo (P, n = 9) intake. The exercise protocol consisted of two bouts (A and B) of maximal contractions, which were interspersed by 15 s rest intervals. Bouts A and B consisted of 15 series of 12 contractions each, separated by a 60 min rest period. During the training period the subjects performed the same exercise protocol twice per day with a 3–5 h rest interval. Blood samples were collected before and after bouts A and B and 24 h after bout B. Knee extension power outputs were similar for P and R for all contraction series. The exercise increased blood lactate and plasma ammonia concentrations ($p < 0.05$), with no significant differences between P and R at any time. After a 6 week washout period in a subgroup of subjects (n = 8) needle biopsy samples were taken from the vastus lateralis (part of the quadriceps thigh muscle) before, immediately after and 24 h after the pre-test. ATP and total adenine nucleotide content were decreased by ~25% and 20% immediately after and 24 h after exercise in both P and R.

It was concluded that 16 g of oral ribose supplementation, at a rate of 4×4 g, taken shortly before and after exercise confers no benefit on muscle ATP recovery or on maximal intermittent exercise performance.

Key points

- The most important metabolic changes during intensive all-out exercise are: (i) depletion of creatine phosphate store; (ii) breakdown of $ATP \Rightarrow AMP \Rightarrow IMP \Rightarrow$ end-products; (iii) increase in muscle and blood lactate; (iv) increase in muscle and blood ammonia; (v) increase in blood xanthine, hypoxanthine, adenine and uric acid.
- After exercise the breakdown products mentioned under (v) might be lost from muscle. This results in a decreased total adenine nucleotide (TAN) content.
- The resynthesis of adenine nucleotides is a slow process so the recovery to a normal level may take up to 3–4 days.
- It has been hypothesized that the oral supply of ribose can lead to a more rapid recovery of the TAN pool after intensive training sessions or competitions but there is no evidence that this is the case in humans.
- The intake of 16 g of ribose was ineffective in raising blood ribose levels significantly to a level that may boost recovery. This dosage was ineffective in enhancing any performance parameter.

CREATINE

Creatine is the most studied ergogenic substance in the last decade. Creatine is not on the doping list of the International Olympic Committee (IOC). Meanwhile, a number of excellent reviews have been published (261–268) that will give the reader abundant and detailed information on this topic. In summary, creatine is synthesized in the human body from the amino acids arginine, methionine and glycine at a rate of 1–2 g/day. Creatine is present in food in small quantities resulting in a daily intake of 0.25–1 g/day. Creatine is mainly present in skeletal muscle, which contains about 95% of the total creatine pool. The total amount in the human body is estimated to be 120 g. In skeletal muscle, creatine plays a role in a number of important metabolic functions.

1. The maintenance of an appropriate level of ATP, through rapid rephosphorylation of ADP from phosphocreatine. This is particularly important during transition from rest to exercise as well as during short term all-out performances.

2. Supporting the creatine phosphate shuttle. Free creatine and phospho-creatine enhance the exchange of high energy phosphate from the site of the mitochondria to the site of the cytosol. Accordingly, ATP produced by oxidative metabolism in the mitochondria may serve the energy requirement in the cytosol during periods of high intensity anaerobic work and related intense glycolysis.

3. Phosphocreatine will help to reduce acidosis in the muscle cells by buffering hydrogen ions. The hydrolysis of phosphocreatine 'consumes' hydrogen ions.

4. The products resulting from the hydrolysis of phosphocreatine—inorganic phosphate and free creatine—may help to regulate the activation of CHO breakdown (glycolytic) processes in the muscle.

5. Creatine may stimulate muscle protein synthesis resulting in an increase in muscle fibre size and lean body mass (304).

Supplementation of creatine monohydrate in amounts of 20–30 g/day for 4–5 days has been shown to increase mean total muscle creatine content by 25–30%. Evidence indicates that the creatine kinase reaction is facilitated in creatine-loaded muscles, due to which the intracellular accumulation of ADP is inhibited during high intensity muscle contractions. Thus, the formation of AMP and IMP is prevented or reduced. Accordingly, the loss of adenine nucleotides is alleviated.

Performance Effects

Creatine supplementation has been observed to result in improved performance in certain conditions. Based upon the functions listed above, the phosphocreatine system is particularly important in conditions that lead to a rapid change in the rate of ATP breakdown and resynthesis. These conditions are determined most importantly by the exercise duration and intensity. Maximal intensity and, accordingly, short duration, will lead to a rapid decline in the muscle phosphocreatine content because the rate of phosphocreatine resynthesis from creatine cannot keep pace with the rate of phosphocreatine hydolysis that is required to maintain high ATP resynthesis from ADP. Consequently, glycogenolysis and glycolysis need to be activated maximally, in order to supply the required high energy phosphate. Thus, the relative contribution of phosphocreatine hydrolysis to supply high energy phosphate compared with the contribution from glycolysis and oxidative processes varies with the intensity and the duration of the exercise. For example, during a very short all-out sprint exercise (6 s) the phosphocreatine system and glycolysis may each contribute about 50% of the total ATP requirement with minor contribution from oxidative phosphorylation processes in the mitochondria, which require substantial time to be upregulated appropriately (260). The phosphocreatine content in muscle will be reduced, but based on the very short duration, not be depleted.

With longer duration, such as 200 m running sprints, 50 m swim sprints, 500 m speed skating, lasting about 20–40 s, glycolysis will be activated maximally and aerobic processes will start to contribute more. Accordingly, the contribution of the phosphocreatine system will be gradually reduced to about 25% (259). Although the quantitative contribution from phosphocreatine is smaller than during a very short sprint, its depletion will last longer due to the longer duration.

With further increasing exercise duration, intensity will fall. Oxidative energy production within the mitochondria from glucose, fatty acids and to a lower extent amino acids, will become appropriate to cover the requirement of energy rich phosphates for ATP resynthesis. In this condition there will not be a substantive decrease in phosphocreatine as seen in short-term exercise. In this condition, the normal phosphocreatine and free creatine content in muscle is abundant to serve the phosphocreatine shuttle appropriately. Glycogenolysis and glycolysis will be lower, resulting in only mildly elevated blood lactate values.

In the light of these observations it is understandable that an elevation of total muscle phosphocreatine and free creatine levels as a result of supplementation may not lead to performance effects in single sprints of short duration or during endurance activities such as a marathon or triathlon. However, creatine supplementation and the resulting increase in total muscle creatine may lead to an improved performance in those conditions where the phosphocreatine store will be depleted, along with a high production of lactate as well as a drop in muscle ATP. Examples of such conditions are sprint and middle distance exercises with a duration of approx. 30–180 s, as well as high intensity intermittent exercise, such as repeated sprints during interval training sessions or interval type sports. Examples of the latter are ice hockey, soccer, rugby, as well as repeated series of lifting weights, as usual in body building training sessions.

Numerous studies have shown that performance capacity during repeated bouts of all-out exercise is improved after a period of creatine supplementation. Based on the currently available data, it is suggested that creatine supplementation may improve sprint and middle distance exercise by supporting a high power output, but is inefficient in affecting endurance exercise performance. The question whether it is possible to enhance the final phase of a cycling race, during which repeated sprints take place, has not been answered by proper research.

Side Effects

There has been some concern about theoretical negative effects of high dosages of creatine ingestion on kidney function because of the high excretion rates. However, thus far no negative side effects have been observed (269). A recent study by Poortmans and Francaux described the effects of long-term creatine ingestion (10 months up to 5 years) on kidney function and concluded that neither short-, medium-term or long-term supplementation induced detrimental effects on the kidney (305). Side effects, apart from an increase in body weight in the range of 1–3 kg, have not been observed. The most comprehensive review on the effects of creatine on performance and body composition can be found in reference 265 and 481.

Key points

- Muscle creatine (TCr) amounts—±120 mmol/kg.dm (±30 mmol/kg.wwt). 1 mmol Cr = ± 131 mg.1 kg muscle contains ±3.5–4 g Cr. A 70 kg male has a muscle mass of ± 30 kg muscle, which contains about 120 g creatine.
- Supplementation may increase muscle creative content to ±160 mmol/kg.dm or 5 g/kg wwt. (25–30% increase).
- The most important creatine functions are supporting the resynthesis of ATP in the phosphagen energy system: $H^+ + PCr + ADP \Leftrightarrow ATP + Cr$. It acts as a temporary energy buffer when ATP degradation is greater than resynthesis. PCR may help to buffer H^+ ions when lactate production is high. The PCR shuttle serves to maintain high ATP levels by transferring energy from mitochondria to cytosol.
- Creatine may stimulate protein synthesis.
- Creatine supplementation increases muscle phosphocreatine (PCr), free creatine (FCr) and total creatine (TCr). However, not all individuals benefit.
- Performance effects: (i) single short-term exercise < 30 s no effects observed; (ii) repeated sprints effective; (iii) anaerobic performance 30–150 s effective/not effective; (iv) endurance exercise, interval type, effective; (v) endurance exercise, continuous, not effective; (vi) strength gain over time is increased with strength training.
- Side effects: body weight increases by a mean of 2–3% after Cr supplementation.
- The decision to supplement or not should depend on the training as well as the competition characteristics of the sports event.

LECITHIN AND CHOLINE

Choline is the precursor for acetylcholine, a neurotransmitter that is of great importance to the central nervous system and for neuromuscular impulse transmission. During stimulation, the choline used for the release of acetylcholine (AC) comes from intracellular sources only. To maintain appropriate levels of free choline, cells take up choline from the blood (418, 477). For the acute production of acetylcholine, choline may also be obtained from the membrane phospholipids (phosphatidyl choline PC) (411, 415–417). During continuous stimulation of the brain, the choline turnover in the brain is higher than the rate at which choline is taken up (418). As a consequence, the free choline concentration will fall and choline from the PC molecules in the cell membrane will be used to compensate for this. This may result in a consistent partial choline depletion (419, 420) as well as a suppressed acetylcholine release, to a greater extent with longer exercise duration (411, 415, 420). In stimulated muscle, a drop in free choline concentration has been shown to induce a fall in AC release and a slowdown of the transmission of the contraction generating impulse (422). Accordingly, experimental animal studies do suggest that a reduction in muscle acetylcholine production, caused by choline unavailability, may contribute to muscle function impairment (423–425). Thus, it appears that a reduced choline availability may affect neuromuscular function as well as the central nervous system and may thus be involved in the aetiology of both physical and mental fatigue, as experienced by endurance athletes.

Does intensive exercise reduce choline availability for the production of acetylcholine? Only a few studies have been done on this subject. During a marathon run it was shown that the serum choline level of a group of participants decreased by 40% (426). Such a reduction is comparable with the effect seen after ingestion of a choline free diet, which affects brain choline levels and neurotransmitter release. In another, non-published (patent related) study (427), it was observed that running 20 miles (32 km) reduced plasma choline by 30% and that choline supplementation resulted in a 5 min faster finishing time. However, full data are not available, and hence it is difficult to judge the validity of these results. In a more recent study it was observed that 2 h of cycling with a speed of 35 km/h, resulted in a fall of plasma choline by 16.9% (412). Looking at the hypothesized relationships between choline availability, AC release and fatigue, it is interesting to speculate on measures to raise the choline availability by means of supplementation. Dietary choline intake has been shown to increase the plasma choline level as well as brain choline levels (410, 414). Lecithin (purified phosphatidyl choline) has been shown to be better absorbed and to induce a longer elevation of plasma choline than the salt choline chloride, which is partially degraded in the gut (430). It is also suggested that the choline bound to PC or lyso-PC is more rapidly taken up into blood (410), and also into the brain, resulting in a larger enhancement of brain acetylcholine levels (429), compared to choline chloride.

Meanwhile several choline supplementation studies have been performed, of which only a few have appeared in the literature. Von Allwörden *et al.* (412) supplemented 10 triathletes with a dose of 0.2 g lecithin (this was 90% purified PC) 1 h before exercise. This resulted in the maintenance of pre-exercise plasma choline levels compared with the placebo group, which showed a decrease of 16.9%. Without exercise, the supplementation resulted in a 26.9% increase of plasma choline. Performance was not evaluated in this study.

One controlled lab study (306) on the effect of ergometer cycling at an intensity of 65% VO_2 max and the effect of choline supplementation did not show performance benefits. It may be that the duration of this work was too short to elicit effects on choline availability for acetylcholine synthesis.

Key points

- Choline is an important substance for the synthesis of the neurotransmitter acetylcholine.
- Acetylcholine has been shown to decrease significantly during intensive endurance exercise. Such a decrease has been suggested to play a role in the development of fatigue.
- Choline or lecithin (phosphatidyl choline) supplementation has been shown to counteract the decrease in plasma choline levels during exercise.
- Currently there are no data to support a beneficial effect of choline or lecithin supplementation on performance indices.
- Effects of these compounds on cell neuronal and muscle cell function during exercise have not been studied to our knowledge.

SINGLE AMINO ACIDS

During the last two decades various amino acid supplements have been suggested to improve performance, mainly because of stimulating hormone secretion, affecting brain metabolism and enhancing mental concentration as well as the drive to perform maximally.

Although most amino acid supplements on the market are targeting strength athletes and bodybuilders, several supplements are also claimed to enhance endurance performance. Research on the performance enhancing effects of such supplements is increasing, but the available data are still limited. The discussion that follows below is based on several recent reviews and investigations (33, 94, 201, 202).

ARGININE AND ORNITHINE

It has been hypothesized that the ingestion of arginine and ornithine may stimulate the release of human growth hormone, which is thought to stimulate muscle growth. Several studies are available on the effect of arginine and/or ornithine supplementation on body composition and/or muscular strength or power. Three of these studies indicated significant increases in lean body mass, an indication of increase in muscle mass and/or decrease in fat mass. However, Williams (202) criticized the experimental methodology of these studies. His recalculation of the data, using appropriate statistical techniques, revealed no significant differences between the supplemented and the placebo treated groups. The other two studies with a correct methodological approach have so far appeared only as abstracts. Both studies reported no significant effect of arginine or a mixture of different amino acids on measures of strength, power, or growth hormone, in well trained weight lifters (81, 192). Currently there are no sound research data to support an ergogenic effect of arginine and ornithine. This may be related to the general ineffectiveness of the amino acids to increase growth hormone levels beyond the range in which normal physiological levels fluctuate daily.

Using a double blind placebo controlled crossover design, Fogelholm *et al.* (307) studied the effects of a 4 day combined L-arginine, L-ornithine and L-lysine supplementation (each 2 g/day, divided into two daily doses) on 24 h level of serum growth hormone and insulin levels in weight lifters. The supplementation was ineffective in changing circulating hormone levels. The authors mention several studies that have shown immediate effects on blood hormone levels of a high dose of amino acids, taken on an empty stomach after overnight fast as well as after infusions.

However, they conclude that the supplements taken by athletes have generally low levels of amino acids and are ineffective in changing hormone secretion. These observations are confirmed by another study of Lambert *et al.* (394).

It is suggested that the very high dosages that are required to cause any significant effect may cause gastric distress (33). Thus far there is little reason to support a beneficial effect of amino acid supplementation on performance.

TRYPTOPHAN AND BRANCHED CHAIN AMINO ACIDS (BCAAs)

Increased levels of tryptophan in the blood may also increase the secretion of growth hormone, but its theoretically most potent ergogenic effect is based upon another mechanism, the formation of serotonin (5-hydroxy-tryptamine) in the brain. Segura and Ventura (170) suggested that this neurotransmitter may improve performance by increasing the tolerance to pain. In support of their hypothesis, they found that 1200 mg of tryptophan consumed in 300 mg doses over a 24 h period increased time to exhaustion and reduced a rating of perceived exertion (RPE) during a treadmill run to exhaustion at an exercise intensity of 80% VO_2max. In contrast to the hypothesis of Ventura, Newsholme (140) has suggested that serotonin may be involved in the development of fatigue. Accordingly, an increased entry of tryptophan from the circulation into the brain may contribute to the development of fatigue. Based on data from animal research, showing that low blood levels of branched chain amino acids (BCAAs) may facilitate the entry of tryptophan into the brain, Newsholme hypothesized that a decrease in serum levels of BCAAs, as often observed during the later stage of endurance exercise, may be a contributing factor to fatigue. Thus, theoretically, BCAA supplements taken during endurance exercise and enhancing the blood BCAA concentration may help to delay the onset of fatigue by decreasing the rate of tryptophan uptake into the brain. However, no good data are available to support this hypothesis.

Blomstrand et al. (388) observed that BCAA supplementation improved running performance during a marathon in slow runners but not in fast runners. The total group showed no significant effect. This study has been criticized for the split in fast and slow runners after the study had been performed. There was no appropriate matching of subjects (388) Vandewalle et al. (184) depleted subjects of muscle glycogen and then let them perform a cycle ergometer ride until exhaustion at an intensity of 75% VO_2max. They reported no beneficial effect of BCAA supplementation. Nor did Galiano and others (67), who supplemented BCAA during prolonged exercise to exhaustion at a workload of 70% VO_2max. Kreider and his associates (99, 122) provided BCAA supplements to five triathletes for 14 days prior to and during a 'half-Ironman' triathlon (2 km swim, 90 km bike, and 21 km run) performed under laboratory conditions. No significant differences were noted between the BCAA and placebo conditions. Other and more recent studies, including a repeated study by Blomstrand showed no effects on performance (388–390, 482). One of the implications of the hypothesis of Newsholme is that the supplementation of tryptophan itself should result in a change of the ration BCAA/tryptophan and accordingly in a drop in performance capacity. However, during a 2 h intensive cycling test tryptophan ingestion did not affect performance (388). Thus, from the available evidence it may be concluded that the validity of Newsholme's hypothesis and the value of BCAA supplementation during exercise have not been substantiated.

BCAAs are known to pass the liver almost exclusively. Accordingly, any protein source that is rich in BCAAs may be an optimal nitrogen supplier for the muscle tissue in periods of recovery when protein synthesis is known to be increased (125, 194). There is however no evidence that BCAAs have any benefit over other protein/amino acid sources in normal healthy

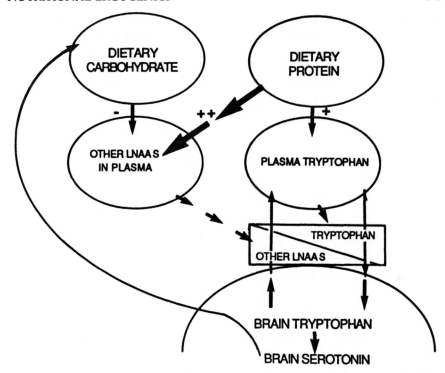

Figure 41 The entry of the amino acid tryptophan, a precursor of serotonin, depends on the ratio of tryptophan/other large neutral amino acids (LNAA's) (leucine, valine, isoleucine) in blood. Carbohydrate ingestion will, via enhanced insulin secretion, lower the LNAA level in blood. This will increase tryptophan entry into the brain. Protein consumption will enhance plasma LNAA's more than plasma tryptophan. This will reduce tryptophan uptake. When serotonin is high, a feedback will lead to reduced CHO consumption. Low serotonin levels may induce CHO snacking. Adapted from R.J. Wurtman, Nutrients that modify brain function, *Scientific American* 1982; 246: 42–51. Copyright © by Scientific American, Inc. All rights reserved

athletes. A recent detailed review on the role of BCAA in exercise metabolism, if any, is given by Wagenmakers (264).

GLUTAMINE

Glutamine is the most abundant amino acid in blood and in the amino acid pool. It has been suggested that an appropriate level of glutamine in the circulation and body fluids is essential for optimal immune competence as well as for protein synthesis (139, 188).

Accordingly, it has been suggested that the decrease in plasma glutamine as observed in endurance athletes may be associated with proneness to infections of the respiratory system. Thus, from a theoretical point of view, glutamine supplementation may be useful to prevent declines in plasma glutamine levels. However, plasma glutamine may also increase as a result of exercise, or remain unchanged, depending on the type, duration and intensity of the exercise.

In endurance athletes who deplete their glycogen stores, the lowest concentration is seen some 2 h post-exercise and it take about 5–7 hours before the concentration is normalized again.

A decline of 10% in plasma glutamine was observed by Parry Billings *et al.* (406) in overtrained athletes compared to non-overtrained athletes and concluded that such a decrease may weaken the immune system. This suggestion seemed to be supported by the observations of Castell who supplied glutamine or placebo immediately after a marathon or ultramarathon and again 2 h later. He observed lower infection rates in the glutamine group during the week post-exercise (407). However, these data should be interpreted with care since the data obtained were only obtained by questionnaire. All symptoms reported, including cough and sore throat, were considered to be indicative of a real infection. There never was an appropriate medical check on these findings. Meanwhile, a number of controlled studies have been done and no effects have been observed that support the findings outlined above. Accordingly, it was concluded in the comprehensive review of Pedersen and Rohde (408) that today there is lack of experimental data to support or reject the hypothesis that glutamine supplementation is of benefit to athletes.

ASPARTATES

The potassium and magnesium salts of aspartate, a non-essential amino acid, have been postulated to improve performance by several mechanisms. The prevailing hypothesis is that aspartates will reduce the accumulation of blood ammonia during exercise. Increases in blood serum ammonia have been correlated with muscular and central fatigue (8, 29, 189). Research data regarding the stimulating effect of aspartates are equivocal. A number of studies have reported no effect. As an example of a well designed study, Maughan and Sadler (116) gave a placebo or 3 g each of potassium and magnesium aspartate to eight subjects 24 h prior to a cycle ergometer ride to exhaustion and reported no beneficial effects. Conversely, an equal number of other studies have documented a positive effect on performance, some reporting greater than 20% improvement in aerobic endurance. For example, Wesson and others (197), who used an appropriate research design, gave a placebo or 10 g of aspartates to subjects over a 1 day period prior to exercise to exhaustion at 75% VO_2max. They reported a significant decrease in serum ammonia levels and a 15% increase in endurance performance. No toxic effects have been reported in studies using these dosages.

It is clear that additional research is needed before recommendations for supplementation can be given, particularly with dosages of 10 g or more because these have been associated with enhanced performance.

L-CARNITINE

L-Carnitine is a compound that primarily facilitates the transport of long chain fatty acids into the mitochondria for their subsequent oxidation in energy production pathways. It has often been suggested that increased carnitine availability may increase the use of fat as substrate for energy production and that this could lead to a sparing of muscle glycogen during exercise. This might increase the time to exhaustion. However, the available data are not supportive of this viewpoint. Data from early studies were inconsistent due to inadequate research design or dosage utilized. For example, several studies from Otto's research group (146, 175) found no effect of 500 mg carnitine taken daily for 4 months on free fatty acid utilization, VO_2max, anaerobic threshold, exercise time to exhaustion, or work output on a cycle ergometer for 60 min. Bucci, however (34) criticized these reports because of the low dosage of L-carnitine ingested. However, several more recent studies, using doses up to 2 g, also reported no effects of carnitine supplementation on fuel utilization at 50% VO_2max, maximal heart rate, anaerobic threshold, VO_2max, muscle carnitine content changes, or exercise time to exhaustion (73, 147, 208, 392, 393). Hultman et al. (308) postulated that the effect claimed by Vecchiet (309), that carnitine supplementation enhances lipid metabolism and reduces lactate formation, cannot be attributed to carnitine. The argumentation was that the bioavailability of orally ingested carnitine is only about 13%. This means that an ingestion of 2 g carnitine will result in an absorbed fraction of only about 0.8–1.6 mmol. When evenly distributed in the muscle, this amount will elevate the muscle carnitine content only by about 30–60 μmol/kg, equivalent to about 1–2% of the total muscle carnitine content. This is clearly not enough to explain any effect. Recently, several reviews have appeared (190, 264). Generally it was concluded that oral L-carnitine does not affect endurance performance or fat metabolism. D,L-Carnitine has been shown to be harmful and should not be taken. L-Carnitine, which is produced in the human body, is relatively harmless when taken orally. However, oral supplementation in healthy human subjects does not lead to increased levels of L-carnitine in muscle, and thus fails to affect muscle energy/fat metabolism in healthy, trained individuals

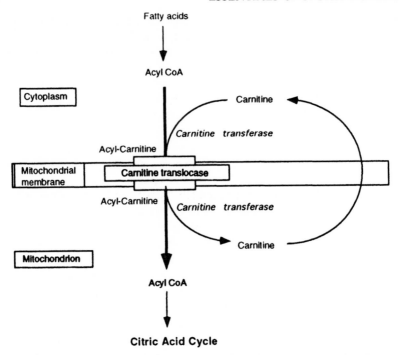

Figure 42 Carnitine is needed to transport long chain fatty acids across the mitochondrial membrane. Activated fatty acids are linked to carnitine through formation of acyl carnitine by the enzyme carnitine transferase. This compound passes the membrane, after which the process is reversed and carnitine is transported back to the cytoplasm. The fatty acid will enter the citric acid cycle to be oxidized. Adapted from R. Laschi, L-carnitine and Ischemia, Foundazione Sigma-Tau, 1987

(190); nor does exercise itself lead to a decrease in total muscle carnitine content (50, 95). A recent review can be found in ref 478.

COQ10 (UBIQUINONE)

CoQ10 is a lipid compound that is present in the mitochondria, particularly in the heart. It has been used therapeutically for the treatment of cardiovascular disease because of its role in oxidative metabolism and as an antioxidant. Because CoQ10 supplementation has induced an increased oxygen uptake and exercise performance in cardiac patients, it has been suggested that it may be effective for performance enhancement of endurance athletes as well. However, there are no data available to support this suggestion. Several recent studies found that CoQ10 supplementation may significantly increase serum CoQ10 levels, compared to a placebo supplementation. However, there were no significant improvements in serum glucose or lactate at submaximal or maximal workloads, cardiovascular function, VO_2max, or endurance performance (21, 159, 210). Demopoulos and others (51) suggested that supplementation with this compound may under certain circumstances actually be hazardous, because it may act as a pro-oxidant when given at high dosages. Accordingly, at high dosage it may induce free radical formation rather than prevent it.

INOSINE

Inosine is a nucleoside. Some of the reported metabolic roles of inosine such as facilitation of ATP (energy rich phosphate) synthesis, effects on muscle glycogen breakdown, and on blood and oxygen supply have been extrapolated to exercise physiology, suggesting that both strength and endurance athletes might benefit from supplementation. Inosine is available either pure or combined with other cofactors, such as CoQ10. Only one study has investigated its effect on endurance parameters. Williams and others (200) used a recommended supplementation protocol (6 g of inosine for 2 days) and reported no significant effects on metabolic parameters and performance. Clearly, additional research is needed to substantiate any hypothetical benefit.

BEE POLLEN

Chemical analysis of bee pollen shows that it is composed of a mixture of vitamins, minerals, amino acids and other nutrients. Although bee pollen does not appear to induce any specific physiologic effect, its theoretical ergogenic effect may be based on the roles that vitamins and minerals are thought to have in exercise metabolism. To test this, highly trained runners were studied and no significant effect on the rate of recovery, as measured by performance in repeated maximal treadmill runs to exhaustion with set recovery periods, could be found (205). Additional well controlled studies have reported no effects on maximal oxygen consumption or other physiological responses to exercise, or on endurance performance (36, 179). Individuals who are allergic to bee pollen may experience an anaphylactic reaction leading to severe health risks (54). These data clearly do not support any good reason to use bee pollen preparations for reasons of improved performance or energy enhancement.

PHOSPHATE SALTS

Phosphorus is an essential mineral that functions in the body as phosphate salt. This is a cofactor or component of several B vitamins, ATP and phosphocreatine, 2,3-DPG (diphosphoglycerate), and an intracellular buffering system. Based on these metabolic roles, it has been suggested that phosphate supplements may improve performance. Some early studies indeed suggested that phosphate salt supplementation was an effective ergogenic for several types of physical performance. Although Boje (20) criticized these studies for design flaws, he indicated that phosphates probably could increase physical performance if consumed in quantities found in the normal diet. Most of the current research, however, has focused on the ability of phosphate salts to enhance oxygen uptake and endurance performance. Some studies supported Boje's observation from over 50 years ago (100, 202). Williams (202) cited four studies, all using appropriate experimental designs and dosages, reporting no beneficial effect. In contrast, four other well designed studies (202) reported significant benefits related to performance. Meanwhile there is a reasonable body of evidence that phosphate loading improves performance by various metabolic and cardiovascular responses in both trained and relatively untrained individuals.

Effects on the exercising athlete are:

- Stimulation of glycolysis by elevating intra- or extracellular phosphate levels (310–312)
- Enhancement of oxidative metabolism and attenuation of the anaerobic threshold (313, 314, 318)
- Favouring the availability of phosphate for oxidative phosphorylation and creatine synthesis (315)
- Promoting oxygen binding in red blood cells (313, 316, 317)
- Improving myocardial and cardiovascular response to exercise (318–320)
- Enhancing buffer capacity (319–321).

Accordingly a number of studies have shown that phosphate supplementation can enhance exercise performance capacity (313, 314, 318, 320–323). Other studies, however, have failed to show positive effects (317, 324, 325). The latter may be due to the type of exercise protocol, the dosage and timing of the phosphate ingested and the training status of the subjects. The majority of studies done in well trained competitive athletes showed positive effects (318, 319, 322, 323). The amount of phosphate ingested is usually several grams/day, often given in three or four dosages, i.e. four times 1 g sodium phosphate/day.

SODIUM BICARBONATE

Sodium bicarbonate is an alkaline salt. Its major function is to control acid–base balance in blood and extracellular fluid. Its proposed role as an ergogenic substance is to buffer the lactic acid, which is produced during high intensity exercise and accumulates in the blood. Such increased buffering may affect the onset of fatigue. Research on the ergogenic effects of sodium bicarbonate has been conducted for over 50 years. In many studies a quantity 0.15–0.40 g (most often 0.30 g) per kilogram body weight was administered 1–3 h prior to an exercise task of maximal intensity and short duration. Such a performance requires mainly muscle glycogen as energy source and results in the production of lactic acid, which is supposed to induce fatigue. Usually these tests were performed to exhaustion and consisted of single bouts of exercise or repeated bouts of sprint exercise with small rest periods in between.

Several reviews regarding the effectiveness of sodium bicarbonate have been published (70, 85, 117). About 50% of the studies reviewed, and being of acceptable quality, have shown a beneficial effect on physical performance and on psychological perceptions of exertion. In one review (85) it was hypothesized that supplementation of an appropriate dosage of sodium bicarbonate appears to have no effect on high intensity performance (30 s or less), or on endurance performance that depends primarily upon oxidative metabolism. However, performance in intense continuous exercise of approximately 1–7.5 min or in repetitive bouts of intense exercise involving short rest intervals may be improved.

Several studies have reported gastrointestinal distress, such as diarrhoea, following ingestion of sodium bicarbonate, while several case studies of gastric wall damage have also been reported. No gastrointestinal problems were reported with sodium citrate, which has the same effects on buffering, as sodium bicarbonate in dosages up to 0.5 g/kg body weight (120).

CAFFEINE

Although caffeine (CAF) is on the IOC doping list it is also a substance that is ingested daily by many athletes. As such it is 'in the grey area between doping and nutrition'. The effects of CAF have been extensively studied and its impact on performance and metabolism make the substance highly interesting to sports physiologists. For this reason CAF will be dealt with in more detail. The following information is largely based on data obtained from Terry Graham, Canada, and is published in great detail in his excellent review on gender differences in the metabolic responses to caffeine (323) and effects of caffeine on tissues (333). Other detailed reviews are given by

Dews (331), Dodd *et al.* (330), Spriet (329) Powers and Dodd (326) and Nehlig and Debry (328).

CAF is the most frequently used stimulant worldwide. Coffee is without doubt the most important source of our daily CAF intake. However, chocolate and soft drinks, especially the designer/energy drinks on the European market, also make significant contributions. Tea also contributes but to a lesser degree. Many countries have a high CAF intake, especially Scandinavia and The Netherlands. The annual, worldwide, consumption of CAF in terms of coffee beans used for the extraction of coffee amounts to 10 kg coffee beans/person/year!

CAF is a naturally occurring compound with the name trimethylxanthine. It is present in a large number of food and drink products as shown in Table 12. There are also three dimethylxanthines (paraxanthine, theophylline and theobromine). These also have a biological activity comparable to CAF. The last two substances are present in chocolate and tea. Additional CAF and theophylline are available as medicinal compounds. As such, they are present in several medical prescriptions (Table 13).

Coffee itself is the watery extract from the coffee bean. It contains a mixture of over 200 chemical compounds, of which CAF is only one (347). The relative presence of all these compounds, with respect to their quantities in the coffee that we drink, depends strongly on the type of coffee beans that are used. Local factors play a role such as the soil and climatic conditions in which the plants have grown. There are two types of

Table 12 Caffeine content of various commercial products

Product	Content	CAF(mg)
Coffee		
—filter	150 ml	115
—instant	150 ml	60
—espresso	60 ml	100
—decaffeinated	150 ml	<1
Tea		
—bag, black	150 ml	30, 50*
—bag, green	150 ml	15, 35*
—leaves	150 ml	20, 30*
Chocolate milk	240 ml	5
Warm chocolate	240 ml	10
Milk chocolate	30 mg	6
Bakery chocolate	30 mg	20
Chocolate cake	1 slice	25
Soft drinks		
—Coca-Cola	360 ml	46
—Pepsi	360 ml	38

*Extraction during respectively 1 and 5 min.

Table 13   Caffeine content of a number of pharmaceutical products

Product	Quantity	CAF (mg)
Pain relievers		
—Exedrin	2 tablets	130
—Anacin	2 tablets	64
—Midol	2 tablets	64
—Cafergot	1 tablet	100
—Fiorinal	1 tablet	40
—Norgesic	1 tablet	30
—Norgesic Forte	1 tablet	60
—Aspirin	1 tablet	0
Laxatives		
—Dristan	2 tablets	32
Stimulants		
—No Doz	1 dose	200
—Vivarin	1 dose	200
Weight loss preparations		
—Dexatrim	1 dose	200

coffee beans, arabica and robusta, which have different organic chemical compositions. The ways in which the beans are roasted and the coffee is prepared (e.g. filter or espresso coffee) will be of influence on the composition of the final drink.

Although statements from sports practice suggest that CAF is the most (abused) stimulant in sports, there is, as far as we know, currently no information about the consumption of coffee and CAF in the various sports events. It is known that CAF is widely used by its participants in endurance events because of its known properties to reduce fatigue. Athletes in sprints and power events use coffee and CAF because they believe that its use will improve reaction time and maximal power output. For the same reason young people use CAF not only to enhance sports performance but also to reduce the need for sleep when enjoying late night parties. With respect to the latter a new market has been developed, especially in Europe, for designer/energy drinks that contain up to 320 mg CAF/l.

The 1993 Canadian survey (334) of the Center for Drugs Free Sports gives important information with respect to the intake of CAF by youngsters. In this survey, teenagers were asked about which aids they had used during the past 12 months with the objective to improve their performance. CAF, used by 27% of the participants, was number 1 on the list.

How Does CAF Work?

A traditional explanation for the ergogenic effect of CAF was that it stimulates the central nervous system, leading to a mobilization of free fatty

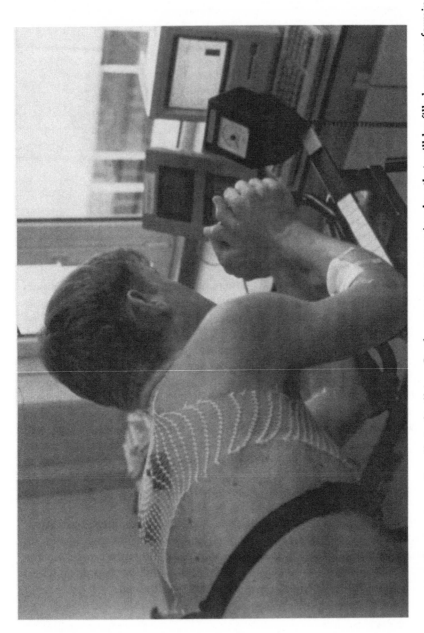

Figure 43 Exercise test to measure the effect of caffeine. On the computer screen is a bar that will be filled upon performing the preset exercise task. When the bar is full, the task is finished. On the back are capsules to sample sweat for the analysis of regional caffeine excretion with sweat

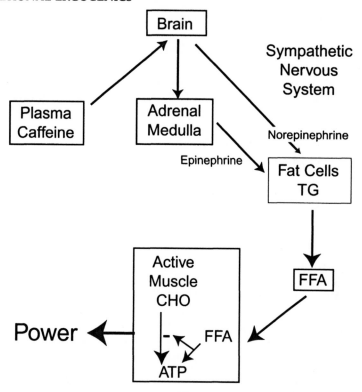

Figure 44   The classical view on how caffeine influences endurance performance. TG = triglycerides, FFA = free fatty acids, CHO = carbohydrate. A minus sign indicates enzyme inhibition (from Graham 493).

acids from adipose tissue. The latter was assumed to enhance the fatty acid uptake by muscle and the subsequent oxidation, in favour of an improved energy production. It was thought that the utilization of muscle glycogen could be reduced by this mechanism.

However, recent studies have led to some doubt about the validity of this theory for the exercising athlete. It was observed that in many studies an increase of the free fatty acid concentration in blood occurred as a result of CAF intake but that this did not lead to an increase in fat oxidation, or to a decrease in glycogen utilization (332). Despite these observations, CAF improved performance in most of these studies. There are additional observations that support the hypothesis that CAF has an effect on performance by mechanisms other than that of modulation of fat and carbohydrate utilization. CAF can improve performance in events that last only 1 to 30 min, as shown in Table 14. Such a short duration is too limited to have a significant influence on the muscle glycogen content. Thus, there must be other factors which underlay the effect of CAF on performance enhancement.

Central Nervous System

It is known that CAF has a stimulating effect on the central nervous system. Therefore, CAF may enhance performance by influencing the processes that determine the stimuli of the neuromotor system. It is also possible that CAF influences the processing of stimuli that enter the central nervous system from the periphery, e.g. by reducing the awareness of feelings related to muscle fatigue.

Other observations show that CAF affects the local processing of nervous stimuli by the muscles. It is suggested that at least a part of the local effects can be explained by increasing the calcium concentration in muscle cells or by reducing the loss of potassium from the cells during the process of repeated contractions. Both mechanisms may influence endurance capacity positively.

Ergonomic Effects and Endurance Performance

Most laboratory trials are concerned with the effect of CAF on the ability to execute a continuous exercise load as long as possible. Table 14 gives a compilation of a substantial number of these studies. There is no doubt about the performance enhancing effects of CAF in endurance events,

Table 14   Studies on the effects of CAF on performance

Author	Protocol	Key results of study
Anselme et al. (335)	Repeated 6 s sprints	CAF improves max power output by 7%
Greer et al. (339)	4 Wingate tests	No effects observed
Collomp et al. (336)	1 Wingate test	No effects observed
Collomp et al. (338)	100 m swimming	Trained swimmers were 1 s faster
Wiles et al. (348)	1500 m running	CAF 4 s faster
MacIntosh and Wright (344)	1500 m swimming	CAF 23 s faster
Cohen (341)	21 km road competition	No effects observed
Berglund and Hemmingsson (333)	Cross country ski 21 km	CAF 3.2% faster
Ivy et al. (340)	Workload performed during 2 hrs cycling	CAF 7.3% more work
Wemple et al. (347)	3 h cycling at 60%; than end sprint as 500 rpm with high resistance (5–6 min)	No effects observed
Kovacs et al. (342)	Cycling time trial, 1 hr	CAF 3.6 min faster
Clinton et al. (343)	2000 m rowing test	CAF 1.2% (range 0.4–1.9%) faster

especially in very long lasting events. There are reports that CAF has a positive effect on 1500 m running and 1500 m swimming, on a 1 h cycling time trial, on 21 km cross country skiing as well as on the capacity to perform work during a 2 h cycle test. Some studies focused on the effects of CAF on short-term high intensity performance. The results of these studies are not as consistent as those of the endurance studies. However, generally it is thought that CAF can have positive effects on high intensity endurance by improving mechanisms that determine the maximal power output at the level of the central nervous system and neuromuscular function. In a recent study, by Maastricht University in The Netherlands, it was shown that the intake of carbohydrate electrolyte solutions with relatively low levels of CAF improved a 1 h time trial performance significantly (342), compared to the no CAF situation. It has also been suggested that CAF may have a positive effect on the quantity of training as well as the quality of the daily training sessions, simply by reducing fatigue and allowing more intense training to be sustained.

Habituation Effects

In sports practice it is often suggested that long-term CAF consumption will reduce sensitivity to CAF. Few studies have focused on the effect of long-term CAF consumption on metabolism. It has been shown that the metabolism of CAF is enhanced over time and that there will be some form of addiction, leading to withdrawal symptoms such as headache and sleeplessness, once CAF intake is stopped. However, no data are available on the effects of long-term consumption on the modulation of sports performance over time. Recently Canadian researchers looked at the effects of CAF withdrawal on endurance performance. No effects were found.

Speed Events

In many sport events it is the combination of the ability to realize a high power output with the ability to develop a maximal speed that will decide the winner. Examples are all middle distance and speed endurance events with durations between 1 and 10 min. Accordingly, a number of studies have reported positive effects of CAF in such events (Table 14). More complex is whether CAF can improve short-term performance in sprint events lasting only a few seconds up to about 1 min. Such performances are influenced by so many factors that it becomes almost impossible to do well controlled studies on the effect of one variable. Today there are no data to support a positive effect of CAF on sprint performance or other events of very short duration such as throwing or jumping events.

Strength and Power Events

Strength athletes often ingest CAF in the belief that this will improve their maximal power or that it will reduce fatigue and improve concentration on days when repeated top performances are required. There are no data to directly support these beliefs.

Timing of Intake

No systematic studies have been done on the effect of the time of CAF intake on performance. The absorption of CAF by the body is relatively fast and is measurable after 15 min. Maximum levels in blood are measured about 60 min after intake. The half-life time of CAF is relatively long. Accordingly, it will take 4–6 h before the CAF level in blood will start to decrease significantly. Thus, in order to obtain a maximal positive effect of CAF on performance, the intake should take place about 1–1.5 h before the time of the performance.

In this respect it is difficult to answer the question about timing in the case of a long lasting performance such as a marathon or triathlon. Recent studies have shown that repeated intake of a CAF containing rehydration beverage (150 mg/l) during a 1 h time trial improved performance significantly (342). Consumption of the same drink during a 4 h event did not lead to unacceptable urinary CAF levels, most probably as a result of sweat induced CAF losses. However, performance capacity was not measured during this long duration study.

Effects on Fluid Balance

At rest, CAF consumption will lead to an increased production of urine. Accordingly, it has often been suggested that endurance athletes who compete in the heat, which might increase the level of dehydration and reduce performance, should not ingest CAF. This cannot be substantiated by scientific data. Several studies have shown that the diuretic effect of CAF is overruled during exercise by the inhibiting effect that intensive exercise itself has on urine production. Thus, CAF has no diuretic effect during exercise (342).

CAF Detection Levels

Sports organizations that check the urine for the presence of CAF as a stimulant only measure the compound trimethylxanthine (caffeine). The liver metabolizes CAF to dimethylxanthines, which also have a stimulating effect. These compounds may be further metabolized and are not measured in the urine test. Urinary CAF excretion amounts to only 1–3% of the

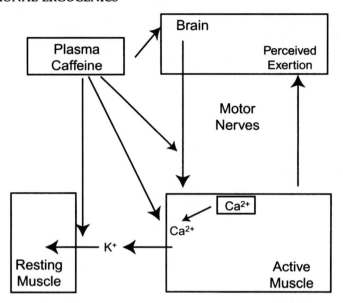

Figure 45 A representation of alternative mechanisms by which caffeine may influence performance (from Graham 493)

amount of CAF ingested. Thus, the urinary CAF content is a poor measure for the determination of abuse of CAF as stimulant.

Another important observation is that there are large inter-individual differences in the excretion of CAF and related compounds. It has been observed that there are athletes who excrete two or three times as much CAF as others, despite a similar CAF intake. As such, the doping level of 12 mg/l as set by the IOC remains a controversial measure.

Athletes who compete in international competitions are advised to test their CAF excretion levels in response to a habitual or desired intake, in order to get information on the level of consumption that is tolerable without running the risk of being positive in a drugs test. Currently no differences between sexes are known that will be of significant influence on the absorption, the metabolism, the distribution or the excretion of CAF.

Side Effects of CAF

CAF ingestion, especially when ingested in larger amounts (>4 mg/kg body weight) may lead to side effects but these are generally mild. CAF can irritate the stomach wall as well as the intestine, which may lead to gastric acid reflux and intestinal motility changes. Occasionally diarrhoea may occur. Studies on the effect of CAF on gastrointestinal function are scarce.

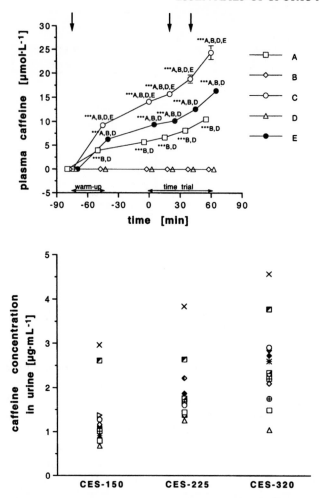

Figures 46 Results of a recent study of Kovacs *et al.* (342) on the effect on performance of caffeine added to a rehydration solution and ingested during exercise. Compared to the water placebo trial, the ingestion of a carbohydrate electrolyte solution (CES) or CES with the addition of resp 150, 225 or 320 mg caffeine/litre improved the performance during a simulated 1 h time trial. Caffeine had a dose response effect. In all cases the urinary caffeine content remained below the IOC limit of 12 mg/litre. Some subjects were consistently lower or higher in their blood and urinary caffeine levels with equal caffeine intakes

Our recent laboratory work on the effects of consumption of a sports drink with 150 mg/l of CAF did not show any effect on gastric acid production, gastric reflux or intestinal motility (270). This will not exclude any possibility that a higher CAF intake, e.g. by CAF capsules, may lead to effects that will disturb the athlete in his performance. Other side effects

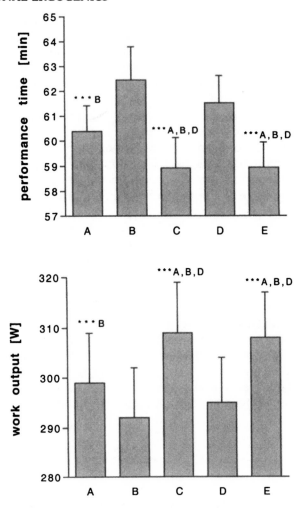

Figure 47   See Figure 46 (previous page) for caption

reported after intake of larger dosages of CAF are a reduced movement coordination, hyperarousal leading to taking wrong decisions and to restlessness/sleeplessness. In these circumstances blood pressure, heart rate and ventilation frequency may be elevated.

SUPPLEMENTS FOR BODYBUILDERS

Supplementation is particularly high among athletes involved in strength sports that are also weight class limited, or are characterized by the athletes' desire to have a high muscle mass/low fat mass, as is the case in body-builders. Several studies have highlighted a very high use of supplements among bodybuilders in the range of 100% among females and 90% among males (362, 363). Brill and Keane (362) studied supplementation patterns of 309 competitive male and female bodybuilders in an age range of 13–70 years. Approximately 94% of all respondents took some kind of supple-ment, comparable to values of 90–100% observed in other studies (362, 363). Popular supplements are amino acid preparations, protein concentrates, protein–carbohydrate combinations, carnitine, inosine, pangamic acid, herbal formulations and micronutrient combinations. Some bodybuilders seem to cycle the use of specific supplements or their combinations depending on their stage of training. In the building phase a special focus is on supplements that are believed to be helpful in improving muscle mass and muscle strength, whereas during the cutting off phase the focus is more on supplements that may help reduce fat mass.

The majority of the bodybuilders (70.3%) reported using supplements in order to meet the extra demands of heavy training and more than 50% to improve training performance and energy levels. The author concludes that many bodybuilders have attached false hopes to these products. Clearly, such false hopes are based on erroneous opinions and information as well as claims that are communicated by the marketing media of supplements companies. Also, coaches and trainers who pick up such information and dispense it as 'their own scientific wisdom' are of significant influence. Education on the physiological aspects of bodybuilding as well as the nutritional efficacy of the marketed supplements is necessary.

Key points

- Substances such as caffeine (guarana), carnitine, aspartates, sodium bicarbonate, bee pollen, specific amino acids, creatine, ribose, choline, etc., have recently received scientific attention due to their possible influence on performance, fatigue and recovery.
- Some of these substances have been shown to be useful for performance enhancement: creatine, phosphate loading, sodium bicarbonate and caffeine.
- Others have clearly been disproved for functional effects at the dosages taken: L-carnitine, bee pollen, BCAA's, tryptophan, inosine, chromium picolinate.
- Some lack the evidence that is required to make benefit statements for the dosage at which the supplements are advised and ingested: ribose, choline, glutamine, arginine, ornithine, CoQ10, aspartates.

- Despite these aspects, supplementation use is widespread, in particular among strength athletes and bodybuilders.
- Caffeine does not act as a diuretic during exercise and enhances performance even at low levels of intake.
- Besides being a nutritional compound, caffeine can also be considered as a drug. It is on the doping list of the IOC and athletes are advised to test for urinary caffeine levels after intake of a standard effective dose before entering competitions.

11 Eating Disorders in Athletes

During the last two decades a substantial number of publications have been attributed to aspects of eating disorders such as anorexia and bulimia as well as to severe dieting practices among athletes involved in sports disciplines in which a low body weight is assumed to be essential for performance (365–373).

Sundgot-Borgen and Corbin (370) reviewed the prevalence of eating disorders in elite female athletes by using questionnaires, interviews and clinical examination in 522 athletes from 35 sports and 448 non-athletic controls in the Norwegian population. A significant higher number of athletes (18%) were found to suffer from eating disorders compared to controls (5%). Especially affected are athletes competing in sports requiring a specific weight or body leanness, such as aesthetic, endurance or weight-class sports.

Pathogenic weight control methods and their frequency are described. Such methods include diet pills, laxatives, diuretics, vomiting, severe fasting, bingeing and excess exercise.

Beals and Manore (380) reviewed the existing literature on the prevalence and consequences of subclinical eating disorders in female athletes. Dietary habits, sports attitudes, body image, energy expenditure and energy intake are dealt with in great detail and the interested reader who wants to learn more about these aspects should consult this review.

The authors conclude that restrictive eating or obsessive weight control behaviour may be self-defeating because severe energy restriction may cause an increase in energy conservation or energy efficiency, which, in itself, may render further attempts at weight loss or weight control less effective. Nevertheless, these athletes compete at a high performance level, which may suggest that a higher efficiency of the body in energy metabolism has taken place (381). A number of recommendations are given for further research on both physiological and psychological aspects. A sport discipline in which the effects of severe diet restriction on performance have been studied is wrestling and aspects of dieting and eating disorders related to the realization and maintenance of a certain body physique have been studied most among female dancers. Both will be discussed in more detail below.

LOW WEIGHT CONCERN: DANCERS

Anorexia and bulimia are common eating disorders in high performance orientated dance companies, especially among dancers having a natural higher body weight compared with those who have a natural ectomorph body type (376–379). Practical tips on how to help the athlete with bulimia can be found in Clark (387). Susan Campbell Sandri (375) reviewed the aspects of body composition and related nutritional problems in dancers. She discusses the basic problem of a culture clash between dancers and nutrition authorities because dancers need safe methods of achieving ultra-lean physique while the recommendations of most nutritionists do not fit with dancers' requirements.

She also states that the favoured ultra-lean body type for female dancers has led to a myriad of effects including delayed menarche, disturbed menstruation patterns and nutritional inadequacies that may lead to negative physiological effects. In extreme cases osteoporosis and chronic tendinitis have been reported. Basic suggestions to be given when advising dancers about nutrition (375) are listed in the box below.

Key points

- If a possible eating disorder is expected, start by making a thorough assessment of the dancer's behaviour and attitudes towards food, including interviews with persons that are of significant impact on the daily life of the dancer.
- Realize that in daily life most persons can hide some excess weight by the way of dressing but that this is impossible for a dancer. Even a small 'overweight' will be visible in the thin dancing costumes.
- Never try to change the dancer's ideals. If you do, you will not be accepted as helpful.
- Realize that drug prescriptions are often rejected in fear that they will affect body weight and may lead to water retention.
- Many athletes consult nutritionists in order to perform better. This is not necessarily the case with dancers. Often their contacts are because others have signalled that the dancer may have a problem requiring counselling. The difference between self-motivation to be counselled versus being brought to a nutritionist by another person is the basis for understanding the dancer's position.
- Dancers often want to lose weight rapidly before major upcoming performances. It should be recognized that fat loss is a slow process, which requires understanding of both the dancer and her instructors. A combination of cutting fat intake, with inclusion of aerobic work sessions, seems to be in place.

- Dancers seem to be more frightened about overweight than about the health impairing effects that severe eating disorders may cause. It is important to educate both dancers and their 'significant others' about the serious effects that eating disorders may have on both dancing performance and health.

OVERWEIGHT CONCERN: WRESTLERS

A concern about some form of overweight is common in most sports events where certain weight categories are in place (Table 1). Wrestling, particularly, has been subject to a great number of studies that have dealt with body composition, eating habits and weight loss regimens. A number of these studies have focused on the impact that rapid weight loss may have on various physical performances parameters. Some excellent reviews and discussions can be found in references 382–386. Basically the following findings have been reported. The primary methods of weight loss are diet manipulation by using well balanced diets, fasting and reduction or elimination of fluid intake. Other measures are aerobics to reduce body fat, dehydration via thermal exposure (sauna, hammam) or exercising in nylon suits or multiple layer clothing. The use of diuretics, laxatives, colon cleaning procedures and very low caloric diet (VLCD) products has also been reported. Some reports mention that a low percentage (up to 4%) of wrestlers are at risk of developing bulimia. Rapid weight loss prior to competition weigh-in procedures, followed by a rapid weight gain as well as the regular repetition of these procedures (weight cycling) have been reviewed by Horswill (382) with respect to their effects on performance and resting metabolic rate. The majority of studies show no effect on anaerobic performance, while aerobic performance generally will be impaired by rapid weight loss procedures. The effects of dehydration and reduced availability of muscle glycogen may cause the latter. Most of the studies that have focused on the effects of a rapid rehydration and weight gain as usual in the competition day show that performance returns to pre-weight loss levels despite the fact that nutritional recovery may still be incomplete.

Key points

- There is no good data to support the hypothesis that weight cycling in wrestlers may reduce resting metabolic rate and in this way be a promoting factor for chronic weight gain.
- It seems prudent that wrestlers need to be educated about long-term strategies to manage optimal weight control and diet counselling.

12 From Theory to Practice

The present chapter is based on a review (476) presented at an international consensus conference entitled 'Advances in training and nutrition for endurance sports: from theory to practice'. The conference took place at the Olympic Training Center 'Papendal' in the Netherlands.

NUTRITION BEFORE EXERCISE

Glycogen Loading

Muscle glycogen depletion and low blood glucose levels have been shown to be major factors in the development of fatigue during endurance exercise. Therefore, it is important to ensure optimal glycogen storage prior to exercise and optimal delivery of carbohydrate (CHO) during exercise.

Of crucial importance in the pre-competition preparation of an endurance athlete is defining the best method to optimize the body's glycogen levels. In the past, Scandinavian researchers introduced a super-compensation diet. Their recommended strategy and diet is as follows. One week prior to an important race, a bout of exhausting endurance exercise is performed in order to deplete the glycogen stores. Over the next 3 days a high fat diet is ingested, ideally with less than 20% of the energy intake as CHO. During the remaining period leading up to the race, the athlete should ingest a high CHO diet with less than 20% of the energy intake being derived from fat. No endurance training should be undertaken during the 6 days prior to the race. This diet training regimen leads to a large increase in the muscle glycogen stores (160–200% greater than the normal resting levels). However, this protocol has serious disadvantages:

- During the high fat, low CHO period, athletes often feel weak and sometimes become unmotivated and lose self-confidence.
- For some athletes it may be difficult to compose a palatable diet consisting of only 20% of the energy from fat (or 20% of the energy from CHO), and a good working knowledge of the CHO and fat content of foods is required.
- The high fat diet may cause gastrointestinal problems such as diarrhoea or abdominal cramping in some athletes.
- Many athletes are reluctant to abstain from training for 3–7 days prior to an important competition.

Because of these disadvantages, a more moderate and more practical dietary training regimen has been evaluated. This regimen also begins with a bout of exhausting exercise 1 week before the race. However, during the 6 days that follow, it is recommended that the dietary CHO intake be progressively increased from the usual 50–55 en% to about 70–75 en%. Over the same period, it is advocated that the training volume is gradually decreased without changing the training intensity: this is called tapering.

This protocol also results in significantly increased glycogen stores (150% of normal resting value) (172), without the side effects so often reported by athletes using the classical regimen. A graphical comparison of both treatments is shown in Figure 48.

How Much CHO is Needed?

CHO intake is often expressed as a percentage of daily intake, but the absolute amount ingested may be more important. For a 70 kg individual, the body CHO stores amount to about 600–700 g (10 g/kg body weight). Ingesting up to 10 g of CHO/kg BW will help to replenish the glycogen stores, but greater amounts than this will not further increase these stores. This is especially important for sports where there are repeated days with very high levels of energy expenditure, such as occur in the Tour de France (165).

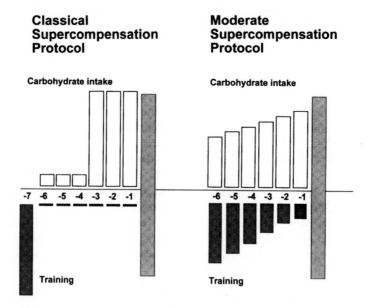

Figure 48 The classical and moderate (tapering) supercompensation protocols as methods to optimize glycogen storage in liver and muscle.

An important point to remember is that CHO is less energy dense than lipids. Consequently, a high CHO diet can be bulky, it is often rich in fibre and may require considerable effort and time to prepare and eat. Examples of CHO rich foods are pasta, potato (well cooked), rice, bread and fruit. During the last 2–3 days before a competition, high fibre foods should be avoided (e.g. green salads/raw vegetables, whole grain bread, unripe bananas, brown rice, muesli) as these may cause gastrointestinal upset.

Pre-race Feedings

During the hours preceding a race it is often recommended that CHO ingestion should be avoided in order to prevent rebound hypoglycaemia. CHO consumption 30–120 min before exercise raises plasma glucose and insulin levels, which stimulate glucose uptake and inhibit fat mobilization and oxidation during exercise. Early studies showed that following a fast, CHO ingestion 45–60 min before an acute bout of exercise could result in a fall in blood glucose concentration soon after exercise had begun. During intense exercise, this was shown to result in hypoglycaemia and a decrease in performance. However, more recent studies that tested subjects in the non-fasted state, which is how most athletes usually enter a competition, did not show a detrimental effect of pre-exercise CHO feeding. These later studies were performed with subjects ingesting different types of CHO meals. Due to the great individual differences in response, however, it is always possible that a certain individual may be prone to exercise-induced rebound hypoglycaemia after consuming a CHO- rich solid or liquid meal.

Based on current information we have established the following guidelines:

Pre-competition Nutritional Guidelines

1. Carbohydrate load using moderate (tapering) supercompensation diet.
2. Ensure a CHO intake of about 600 g/day during the 3 days before the race. Intake of more than this amount may not further increase glycogen storage and is therefore not necessary.
3. Drink plenty of fluids during the days before the race, to ensure that you are well hydrated at the start of the event. If substantial sweat losses are to be expected during the race (see the section on sports drinks below) add a small amount of sodium chloride (about the tip of a teaspoon of table salt per litre) to the drinks.
4. Avoid foods with a high dietary fibre content during the days before the competition to prevent gastrointestinal problems.
5. Eat a CHO rich pre-event meal 2–4 h before a race to ensure adequate levels of glycogen in the liver. Before races of short duration ingest easily digestible CHO foods or energy drinks. Before races of long duration eat

semi-solid or solid food such as energy bars or bread, and keep the intake of fat and protein low.

6. Some individuals may develop rebound hypoglycaemia following a high CHO meal or drink before a race. These individuals should delay eating carbohydrates until the warm up, or within a few minutes of the start of the race. Top athletes should undergo a CHO tolerance exercise test to define their individual response to high CHO intake.

NUTRITION DURING EXERCISE

Carbohydrate ingestion during exercise has been shown to improve exercise performance in events lasting 60 min or longer by maintaining high plasma glucose levels and high CHO oxidation rates. From numerous studies, it appears that most of the soluble carbohydrates are oxidized at similar rates (i.e. glucose, maltose, sucrose, glucose polymers and dispersable starch). The exceptions are fructose, galactose and insoluble starch, which are oxidized at slightly slower rates. Interestingly, however, is the finding from one particular study that when 50 g of fructose and 50 g of glucose were ingested together, during exercise, the cumulative amount of CHO oxidized was 21% greater compared with the ingestion of 100 g of glucose (409).

The amount of CHO ingested is important for its contribution to energy expenditure and sparing of liver glycogen. However, the oxidation of exogenous CHO does not exceed 1.0–1.1 g/min, even when much greater quantities are ingested. This observation suggests that the maximum CHO intake during exercise should not exceed 60 g/h. Nowadays, CHO electrolyte drinks and energy bars, which are promoted to give rapid provision of CHO and fluid, are the most common food supplements in endurance sports. Untrained individuals may benefit as much from the CHO fluid supply as trained athletes.

Optimally, athletes should ingest a CHO electrolyte drink throughout exercise. It has recently been shown that ingestion of CHO throughout exercise improves performance more than when an identical amount of CHO is consumed late in the exercise period.

Sports Drinks

The ideal nutritional strategy during exercise should:

- provide sufficient CHO to maintain blood glucose levels and CHO oxidation
- provide water and electrolytes to prevent fluid imbalance
- not cause any gastro-intestinal discomfort
- taste good.

Optimal CHO sources have been mentioned previously and are also discussed in the section on post-exercise recovery below.

The effectiveness of a sports drink in supporting fluid balance depends on a number of factors of which CHO and sodium content, and osmolality are very important. The ideal sports drink for CHO and fluid replacement should have a relatively low CHO content of between 40 and 80 g/l, have an osmolality which is moderately hypotonic to isotonic, and have a sodium content of between 400 and 1200 mg/l. Individual sweat loss can be estimated from weight loss. By regularly monitoring nude body weight before and after training sessions and competitions, it is possible to predict an individual's fluid loss in a certain race under most environmental conditions. Weight loss will be due not only to fluid loss but also to glycogen and fat oxidation; for example, over 90 min of exercise 100–250 g of substrate may be oxidized. However, since the main limitation to maintaining fluid balance appears to be the volume of beverage that can be tolerated in the gastrointestinal tract, in most situations it is advisable to drink as much as possible. Completely restoring sweat losses by fluid consumption may not always be possible because these losses may exceed 2 l/h, and ingestion of such amounts usually cannot be accepted by the gastrointestinal tract. Therefore, the volume of drink that can be tolerated by the intestine usually limits fluid and CHO consumption. This highlights the importance of making 'drinking during exercise' a part of the regular training programme.

The palatability of a drink is very important because it stimulates consumption and hence increases the intake of fluid and CHO. In addition, the taste and flavour of a drink may also influence the rate of gastric emptying. Flavours and aromas, which are perceived as being unpleasant, may slow gastric emptying and may even cause nausea.

GUIDELINES FOR NUTRITION DURING EXERCISE

1. During intense exercise lasting >45 min a CHO drink should be ingested. This may improve performance by reducing/delaying fatigue.
2. Consume 60 g of CHO per hour of exercise. This can be optimally combined with fluid in quantities related to needs determined by environmental conditions, individual sweat rates and gastrointestinal tolerance.
3. During exercise of < 45 min duration there appears to be little need to consume CHO.
4. The type of soluble CHO (glucose, sucrose, glucose polymer, etc.) ingested does not appear to make much difference when ingested in low to moderate quantities; fructose and galactose are less

effective. However, a combination of fructose and glucose may have physiological benefits. Insoluble CHO sources are relatively slowly absorbed and oxidized, and are therefore not recommended for high intensity events.

5. Athletes should consume beverages containing CHO throughout exercise, rather than water during the early part of an exercise bout followed by CHO beverages at the later stages of the exercise.

6. Avoid drinks which have extremely high CHO contents (>20%) and those with a high osmolality (>500 mosmol/kg) because fluid delivery will be hampered and gastrointestinal problems may occur.

7. Try to predict the fluid loss during endurance events of >90 min. The volume of fluid to be ingested should in principle at least equal the predicted fluid loss. While exercising in warm weather with low humidity, athletes have to drink more to replace sweat loss and the drinks can be diluted. During events in cold weather, athletes require less fluid volume to maintain fluid balance but will still require the CHO to maintain blood glucose levels, therefore the CHO content of the drinks can be more concentrated.

8. Large volumes of a drink stimulate gastric emptying more than small volumes. Therefore, we recommend that athletes ingest a fluid volume of 6–8 ml/kg BW, 3–5 min prior to the start to 'prime' the stomach, followed by smaller amounts (2–3 ml/kg BW) every ~15–20 min.

9. The volume of fluid that athletes can ingest is usually limited. Athletes should practise drinking while exercising as training can increase the volume that the gastrointestinal tract will tolerate.

10. After drinking a large quantity, the stomach may feel empty and uncomfortable. If this occurs it may be wise to eat some easily digested solid food. During long, low intensity competitions solid food can be eaten in the early stages of the event.

11. Fibre and protein content, and high CHO concentration and osmolality have been associated with the development of gastrointestinal symptoms during exercise, and thus should be avoided.

Banana Tips

Banana is said to be very high in magnesium. Is banana a better carbohydrate source for athletes (Table 15)? Banana is a popular carbohydrate snack among endurance athletes. As well as water, the banana has a high starch content. However, unripe bananas (green or yellow skin with green point) are to a large extent indigestible (they have a high content of resistant

Table 15 Modification in carbohydrate composition during banana ripening

	0 days	2 days	4 days	6 days	8 days
Appearance	Green with a bit of yellow	Yellow with a bit of green	Yellow	Yellow with some black spots	Yellow with many black spots
Carbohydrates (g/100 g)	28	29	28	27	26
Resistant starch (%)	82	41	26	9	3
Sugars (%)	7	48	63	81	88
Rest (%)	11	11	11	10	9
Digestibility	bad	moderate	good	good	very good

starch). This means that the starch cannot be digested by intestinal enzymes and that it arrives in the colon undigested. Subsequently it will be fermented by bacteria, leading to substantial gas production as well as to the formation of bioacids such as short chain fatty acids and lactate. During exercise the gas formation may cause an unpleasant bloating. Thus, athletes should only consume really ripe bananas (yellow with small black spots) during exercise.

NUTRITION AFTER EXERCISE

Quick recovery is an extremely important aspect of training and frequent competitions. During repeated days of heavy training it is important to recover quickly in order to maintain the level and volume of training required to improve performance. Dietary measures have been shown to significantly influence recovery. The restoration of muscle glycogen stores and renewal of fluid balance after heavy training or competition are probably the two most important factors determining the time required to recover. The rate at which glycogen can be formed (synthesized) is dependent on several factors:

1. The amount of CHO ingested
2. The type of CHO
3. The timing of CHO ingestion after exercise.

The Amount of CHO Ingested

The quantity of CHO is by far the most important factor determining the rate of glycogen resynthesis. It appears that an intake of 50 g of CHO ingested every 2 h doubles the muscle glycogen resynthesis rate compared with half that amount of CHO consumed every 2 h. When more than 50 g was ingested (100–225 g) in the same period, there was no further increase in muscle glycogen storage. Therefore, 50 g of CHO every 2 h (or 25 g/h) appears to result in the maximum rate of post-exercise muscle glycogen resynthesis. Frequent small meals do not appear to give any advantage compared to eating a few large meals.

Interestingly, the addition of easily digestible protein sources to the CHO may further increase glycogen resynthesis rates.

The Type of CHO

To ensure full restoration of the glycogen stores after exercise, the CHO sources needed must be easily digested and absorbed. The rate of absorption of each CHO source is reflected in its glycaemic index. In Tables 16–18 foods are listed with high, moderate and low glycaemic indexes respectively.

Foods with moderate to high glycaemic indexes enter the bloodstream relatively rapidly resulting in similar rates of glycogen storage; foods with a low glycaemic index enter the bloodstream more slowly and result in lower

Table 16 CHO containing foods with a high glycaemic index (foods listed as eaten). Table adapted from Pennington. Food values of portions commonly used. Harper & Row publishers. 15th edition New York

Food group	Food item giving 50 g CHO	Serving size (g or ml)	Fat per serving (g)
Cereals	White bread	201 g	2
	Wholemeal bread	120 g	3
	Rye bread	104 g	4
	Rice (whole grain)	196 g	1
	Rice (white)	169 g	0.5
Breakfast cereals	Corn flakes	59 g	1
	Muesli	76 g	6
Biscuit	Whole wheat biscuits	76 g	16
Vegetables	Sweet corn	219 g	5
	Broad beans	704 g	4
	Potato (instant)	310 g	0.5
	Potato (boiled)	254 g	trace
Fruit	Raisins	78 g	trace
	Banana	260 g	1
Sugars	Glucose	50 g	0
	Maltose	50 g	0
	Honey	67 g	3
	Sucrose	50 g	0
	Corn syrup	63 g	0
Beverages	6% sucrose solution	833 ml	0
	7.5% maltodextrin and sugar	666 ml	0
	10% carbonated soft drink	500 ml	0
	20% maltodextrin	250 ml	0

Table 17 CHO containing foods with a moderate glycaemic index (foods listed as eaten). Table adapted from Pennington. Food values of portions commonly used. Harper & Row publishers. 15th edition New York

Food group	Food item giving 50 g CHO	Serving size (g)	Fat per serving (g)
Cereals	Spaghetti/macaroni	198	1
	Noodles (oriental)	370	14
Breakfast cereals	Wheat bran nuggets	232	13
	Oatmeal	69	1
Biscuits	Oatmeal biscuits	79	15
	Sponge cake	93	6
Fruit	Grapes (black)	323	trace
	Grapes (green)	310	trace
	Orange	420–600	trace

Table 18 CHO containing foods with a low glycaemic index (foods listed as eaten). Table adapted from Pennington. Food values of portions commonly used. Harper & Row publishers. 15th edition New York

Food group	Food item giving 50 g CHO	Serving size (g or ml)	Fat per serving (g)
Fruits	Apples	400 g	trace
	Apple sauce	290 g	trace
	Cherries	420 g	trace
	Dates (dried)	78 g	trace
	Peaches	450–550 g	trace
	Plums	400–550 g	trace
Legumes	Haricot beans	301 g	2
	Red lentils	294 g	2
Sugars	Fructose	50 g	0
Dairy products	Ice cream	202 g	13
	Milk (whole)	1.1 l	40
	Milk (skim)	1.0 l	1
	Yogurt (plain, low fat)	800 g	8
	Yogurt (fruit, low fat)	280 g	3
Soup	Tomato soup	734 ml	6

rates of glycogen resynthesis. Therefore, it is recommended that low glycaemic index foods should not constitute the main source of CHO intake after exercise *when rapid recovery is required.*

Timing of CHO Intake

During the first hours following exercise, glycogen resynthesis proceeds at a rate which is somewhat faster than that which occurs later. Therefore, where recovery times are necessarily short, CHO intake should take place immediately after exercise. Although this can maximize the rate of glycogen resynthesis in the early phase of recovery, the full process of glycogen storage still takes a considerable time. Depending on the degree of glycogen depletion and type of meals consumed, it may take 10–36 h to restore the body's carbohydrate stores to pre-exercise levels. Therefore it is impossible to perform two or more strenuous workouts per day without depleting the initial glycogen stores. Even when CHO intake between training bouts or competitions is high, the muscle glycogen levels will be suboptimal if the next activity is started within 8–16 h of completing the first activity. The rate at which fluid balance can be restored depends on: (i) the quantity of fluid consumed; and (ii) the composition of the fluid, especially the CHO-sodium content.

Recent studies have shown that post-exercise fluid retention is only about 50% of the volume ingested when low sodium beverages are consumed. Drinks such as most tap and mineral waters and fruit juices have insufficient sodium content to be effective post-exercise rehydration fluids. After the consumption of well formulated CHO-electrolyte solutions containing 40–80 g CHO and 600–1200 mg sodium per litre, the amount of ingested fluid retained may be as high as 70–80% of the intake volume. From these studies it can be concluded that in order to restore fluid balance, the volume of drink consumed post-exercise must be considerably higher (150–200%) of the amount of water lost as sweat.

Practical Considerations

Usually appetite is suppressed after strenuous exercise and there is a preference to drink rather than to eat solid food. Therefore, beverages which contain high glycaemic index CHO sources in sufficient quantities (≥6 g/100 ml) should be made available.

If preferred, the athlete may also ingest easily digestible solid CHO-rich foods such as ripe bananas, rice cakes or sweets. When the desire for normal meals returns, approximately 10 g of CHO/kg BW of moderate to high glycaemic index CHO sources should be eaten within 24 h. This can be easily achieved by consuming foods that are low in fat. For practical reasons a certain amount of low glycaemic index CHO cannot be excluded from the diet.

As the time spent sleeping restricts the number of hours available for eating, it is recommended that before going to sleep, an amount of CHO is eaten that is sufficient to supply the required 25 g of CHO/h (e.g. 250 g for a 10 h period).

POST-EXERCISE NUTRITIONAL GUIDELINES

1. To maximize glycogen storage, it is recommended that during the first 2 h after exercise 100 g of CHO be ingested in the form of liquids or easily digestible solid or semi-solid food. Thereafter 25 g/h is recommended. In total about 10 g of CHO/kg BW should be eaten within 24 h; two-thirds of this amount should preferably be high glycaemic index foods.
2. It is recommended that CHO sources with a moderate to high glycaemic index are eaten (see Tables 16–18).
3. The addition of protein to the CHO consumed during the first hours post-exercise may stimulate glycogen recovery rates.
4. There is no benefit in consuming amino acids or mixtures of amino acids.
5. The addition of 600–1200 mg of sodium to post-exercise rehydration beverages improves fluid retention and the recovery of fluid balance.

DEVELOPMENTS IN 'NUTRITIONAL TRAINING'

1. Athletes should consider their specific energy and fluid demands of training and competition and, in the face of their habitual diet, adopt a pattern of 'nutritional periodization' before major events.
2. Nutritional periodization for endurance and particularly ultra-endurance (>4 h) events should aim to increase the contribution of fat to energy metabolism, and thus spare the body's CHO stores.
3. Individuals should consider trying to improve their performance times for ultra-endurance events by training for most of the year on a 'normal sports adapted diet'. This should be followed by undertaking a 7–10 day period of fat adaptation, prior to the final CHO-loading period carried out over the last 2–3 days before a competition.
4. The effects of adapting to a high fat diet or to medium chain triglyceride ingestion, on energy metabolism and performance during ultra-endurance events requires further research before recommendations can be made.
5. Athletes should practise 'volume drinking' while exercising during training as this may result in a substantial increase in the amount of fluid that can be tolerated during intense competition, without causing gastrointestinal upset.

ESSENTIALS OF DRINKING DURING SPORTS

Timing of Fluid Intake

To maintain optimal sports performance it is recommended to avoid dehydration caused by: (i) large sweat losses; and (ii) carbohydrate the breakdown of the carbohydrate stores in the body for energy. Dehydration can be avoided by ingesting fluid in approximate the amount of body weight that is lost during ᵣbohydrate depletion can be delayed by ingesting carbohydrate ᴀt can be used for fuel delivery to the muscle. This will make it leᴄ ᵣssary to break down local carbohydrate stores or will replenish them ᵢₙ they have been emptied. The main result of such a supply of fluid combined with carbohydrate will be a delay in the development of fatigue and an overall improvement in performance.

Once we know this background information it is easy to understand that athletes should drink carbohydrate-containing fluids in all circumstances where sweat loss and/or carbohydrate breakdown is *large and performance limiting*. In general, this is the case in all circumstances where exercise intensity is high and exercise duration lasts more than 45 min. With shorter duration, in most cases the carbohydrate stores will not be limiting and sweat loss will not be large enough to impair performance or to threaten health. Additionally, with exercise of short duration, the high intensity and the related hyperventilation will make it practically impossible to drink. A good example here is a 10 km run on the track.

What is Right?

It is often stated that the carbohydrate store of the body is sufficient to exercise for 90 min. Therefore, there is no need to ingest carbohydrate for exercise of shorter duration.

The figure of 90 min is most often based on two different observations. The first comes from studies carried out on endurance athletes. If, for example, long distance runners, cyclists or triathletes compete in multi-hour events, the exercise intensity will be such that the rate of carbohydrate breakdown is not maximal. In these cases signs of carbohydrate depletion first become apparent after 1.5 h of exercise.

The second observation is based on the total amount of carbohydrate stored in the body. In untrained individuals this is about 400 g. In trained athletes this may be increased to about 500 g.

If we take the latter figure and multiply it by 4 (the calorific value for each gram of carbohydrate), we get 2000 kcal. This amount of energy is sufficient to exercise for 1.5 h at high exercise intensity.

However, here there is a misunderstanding of physiology because it assumes that all the stored carbohydrate in the body will be available to the muscles that are used in the execution of the sports event. This, is not the case! If leg muscles perform most of the work and the carbohydrate stores in these muscles become depleted, it is not possible to bring carbohydrate from, for example, the arm muscles to the legs. Thus, at high intensity exercise, it is the local muscle carbohydrate stores that limit performance.

Studies on middle distance runners who perform interval training sessions or tempo runs, as well as studies performed on athletes involved in interval-type events (multiple sprint) such as soccer and ice hockey, show that the rate of glycogen breakdown in the most active muscle can be so high that its carbohydrate depletion can occur in less than 45 min. In one particular study it was observed that a 30 s sprint reduced the glycogen content in active muscles by 25%.

For this reason it is recommended to ingest carbohydrate-containing fluids for all events lasting longer than 45 min and characterized by high exercise intensity. A good example of this comes from a Scandinavian study which showed that soccer players ingesting extra carbohydrate played significantly better, as measured by the total distance covered and the amount of sprint work performed, in the second half of a competition.

What is the Optimal Composition of a Sports Drink?

If we take the information given above as a starting point we can say that any drink taken to optimize performance should supply fluid and carbohydrate to the circulation and the cells at a high rate. This can only be achieved when the drink is emptied from the stomach and absorbed in the gut at a high rate. One should know that drinking in itself does not guarantee that the drink is available to the organism. The stomach is a kind of holding tank. After the fluid is delivered to the gut, it must first be absorbed before it can circulate through the body and reach the muscle cells. Thus, the optimal composition of a drink is determined by the factors that affect the rate of gastric emptying as well as the rate of absorption. These factors are especially the carbohydrate/energy content, the mineral content and the osmolarity of the drink.

What Does Osmolarity Mean?

Osmolarity is a measure of the osmotic pressure exerted by a fluid across a biological membrane. If two solutions have the same effective osmotic pressure, these solutions are said to be isotonic. If two solutions differ in osmotic pressure, the one with the higher osmotic pressure is said to be hypertonic, compared to the solution with the lower osmolarity. The latter is said to be hypotonic.

Osmolarity is generally determined by the number of osmotically active particles which are 'in solution', or dissolved. Osmolarity can be measured in a laboratory by a method called freezing point depression. More particles 'in solution' lead to a longer time needed to reach the freezing point and thus to a higher osmotic value.

In a biological system osmolarity influences the way in which a fluid will shift from one compartment to the other. For example, in the gut there is always a shift of fluid in two directions. One, called absorption, is from the gut lumen into the gut cells and then into the blood. The other, from the blood and the gut cells into the gut lumen, is called secretion. As long as absorption is greater than secretion, there is net absorption. However, when secretion is larger than absorption there is net secretion. The latter occurs in diseases or disorders that result in diarrhoea.

Hypertonic fluids are known to increase secretion. Thus, hypertonic fluids decrease the rate of net fluid absorption. Therefore, hypertonic drinks should not be given in a situation where a high rate of fluid uptake is required. Unfortunately there are many drink producers who do not mention the osmolarity on the product label. Many of the commonly consumed refreshment drinks, such as fruit juices and soft drinks, have an osmolarity higher than 600 mosmol/kg.

Which Drink Composition Leads to a Rapid Fluid Availability?

Drinks, which contain 40–80 g of carbohydrate per litre and have an osmolarity below 400 mosmol/kg, preferably hypotonic, are generally found to be effective for sport events, in terms of a rapid fluid supply. The rate at which carbohydrate ingested during exercise can be used by the body amounts to approximately 0.5–1.1 g/min, depending on the exercise intensity and the degree of carbohydrate depletion in the body. The amount of fluid which can be maximally consumed in endurance events amounts to about 400–800 ml/h, depending on the body size of the individual and the type of endurance event. Accordingly, in order to combine CHO with water effectively, it is recommended to ingest drinks containing 40–80 g carbohydrate per litre to obtain a sufficient carbohydrate supply.

What to Consume When Rapid Fluid Availability is Not Important?

There is no need to consume a lot of fluid when, despite high exercise intensity, sweat loss is small, for example when cool air surrounding the athlete significantly cools the body. In this condition it is clear that dehydration will be small and will not limit performance. The main limiting factor in this case is carbohydrate availability. Recalling the body's utilization of oral carbohydrate discussed above, with a mean value of about 0.8 g/min, the athlete should ingest enough carbohydrate-containing fluid, to satisfy this need.

In order to avoid over-consumption of fluid, and the consequent need to urinate during exercise, the athlete is recommended to consume a more concentrated drink, i.e. containing about 130–150 g/l. Such an energy-rich drink can have a low osmolarity when the carbohydrate used is of a complex nature, such as maltodextrins (glucose polymers) or dispersable starch. Based on the high carbohydrate/energy content, such a drink will be emptied from the stomach more slowly than a less concentrated drink. This will reduce the delivery of fluid into the gut, resulting in a slower rate of fluid absorption and delivery to the circulation. However, this is not of concern since fluid availability is not a limiting factor in this situation. It is much more important that the delivery of carbohydrate to the gut is increased, resulting in larger carbohydrate absorption. The latter will influence carbohydrate availability strongly and thus affect performance.

By consuming more concentrated drinks the athlete can reduce fluid intake to 300–500 ml/h without compromising carbohydrate intake to a level of less than 0.8 g/min. As with fluid/energy replacement drinks carbohydrate/energy drinks should not be of high osmolarity, i.e. not >400 mosm/kg, preferably hypotonic to isotonic.

What is the Best Drink Temperature?

When the competition takes place in warm conditions it is best to consume chilled drinks. In general most people like cool drinks more because they taste better. Some athletes like very cool drinks, i.e. 5–10 °C. The cooler a drink is the more it will take up heat from the body. However, tolerance to consume cold drinks, especially in larger quantities, without getting gastrointestinal upset is individually different. Therefore, it is strongly recommended to find the most suitable drink temperature in training sessions. An early study indicated that cold drinks are emptied from the stomach at a higher rate than warm drinks. However, more recent research has shown that this is not the case. May drinks to be taken during exercise in the cold be warm? The answer to this question is yes! Since warm drinks (at a temperature at which a drink can rapidly be consumed) do not delay gastric emptying, do not affect thermoregulation negatively and may bring a psychological benefit in a fatigued athlete in a cold climate, it is recommended that drinks be warm if the athletes wish.

Do Women Need to Drink More Than Men?

No! In general women lose less sweat than men do. It is thought that this is due to more economical sweating. An additional factor is that women simply have a lower body weight than most men, producing less sweat at the same absolute workload. Women should drink enough to compensate for their sweat losses bearing in mind that these are less than in men. The best way to find out the individual fluid needs is frequently to take body weight before and after exercise and to correct for fluid intake. Any body weight loss of >1 kg means a real fluid deficit.

How to Drink During a Soccer Competition

It is not allowed to consume drinks during a soccer match. Knowing the effects of dehydration on body function and performance, this may sound strange. Until now the FIFA has not established adequate drinking rules. It is therefore recommended to consume 300–500 ml immediately before the start of a competition, and further, to drink at moments when the referee allows drink intake, despite the official rules. During the break the players should again consume 300–500 ml.

In warm conditions the drink should be low in carbohydrate (60–80 g/l). In cold conditions it is recommended to consume a carbohydrate/energy drink in smaller quantities (200–300 ml). This drink may be warm.

13   A Brief Outline of Metabolism

GLYCOGEN

Glycogen is a glucose polymer. It is a storage form of glucose in human muscle and liver, roughly comparable to the storage of glucose in plant starches such as potato, rice, grains and banana. Glycogen is synthesized or broken down by different enzymes within the cytoplasm. When synthesized, glucose is phosphorylated to glucose-1-phosphate. Glucose-1-phosphate is converted to uridine diphosphate (UDP) glucose, which is built into glycogen by the action of the enzyme glycogen synthetase. When the amount of glucose is insufficient, glycogen is broken down by action of the enzyme glycogen phosphorylase. Glycogen is mainly synthesized in periods when the amount of glucose present in cells exceeds the amount required for energy production. Glycogen metabolism in the liver regulates the blood glucose level. After meals, glucose and fructose are taken up by the liver, leading to liver glycogen storage. During the night or during fasting, liver glycogen will be broken down to maintain a normal blood glucose level. Muscle glycogen is primarily meant to be a rapid energy source, to be available in a situation of sudden intensive muscular work.

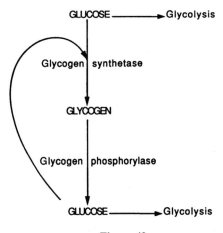

Figure 49

GLYCOGEN AND GLUCOSE METABOLISM

Synthesis or degradation of glycogen in the liver and the muscle is regulated by many factors. Synthesis will normally take place if the supply of glucose 'building units' exceeds the need of glucose for energy production, i.e. if the amount of glucose within the cell increases. This situation occurs after meals, when during a state of physical relaxation the digestion and absorption of carbohydrate lead to increased blood glucose levels in a hormonal milieu which favours synthesis. Thus, insulin will be high, glucagon and stress hormones will be low. In this situation the cells will take up glucose and the enzyme glycogen synthetase will be activated (+), whereas glycogen phosphorylase will be inhibited (−).

In the case of a rapid energy requirement, a number of signals of central nervous system and hormonal origin will cause stress hormones and glucagon to be increased and insulin to be decreased. The enzyme glycogen synthethase will be inhibited (−) and the degrading enzyme glycogen phosphorylase will be activated (+), resulting in the liberation of glucose-1-phosphate from the glycogen pool.

When used in energy production, glucose enters the glycolytic pathway in which it is converted in a number of steps to pyruvate. Depending on the quantitative need for energy, pyruvate is either largely converted to lactic acid, which is the case during an intensive stimulation of glycolysis, such as during supramaximal sport activities (0.5–3 min duration), or pyruvate is taken up in the oxidative energy pathway, the citric acid (Krebs) cycle, which mainly takes place during endurance events. The pathway of glucose → lactic

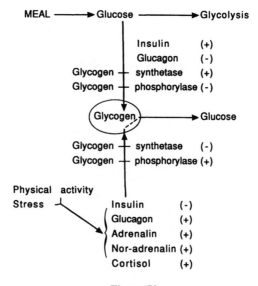

Figure 50

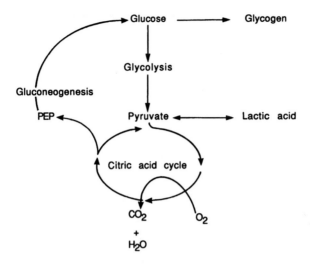

Figure 51

acid is reversible, which means that a high lactic acid content in the blood after intensive sport activity can be lowered by the conversion of lactate back to glucose. This takes place via a different metabolic pathway called gluconeogenesis. Lactate can also be oxidized or converted to fat. During the conversion of glucose to lactate, 2 moles of adenosine triphosphate (ATP) are produced per mole of glucose. During complete oxidation of glucose within the citric acid cycle, pyruvate is converted to water and carbon dioxide and a total of 36 moles of ATP are produced.

ADIPOSE TISSUE/TRIACYLGLYCEROL

Fatty acids are stored in the body as triacylglycerols (triglyceride) in fat cells which make up the adipose tissue. Fat is also stored in muscle tissue in the form of triglyceride, present in small intramuscular fat droplets. After a meal, fat is absorbed and circulates in the blood as triglycerides in the form of circulating lipid particles (HDL, VLDL, LDL, chylomicrons) or as free fatty acids bound to albumin, called non-esterified fatty acids (NEFA). As with glycogen, the synthesis of fat or its degradation depend on the concentration of the 'building blocks', in this case fatty acids. This concentration is determined mainly by uptake of free fatty acids in and from triacylglycerols and their rate of utilization for energy metabolism.

Thus, when energy production is low, the supply of fatty acids after a meal will lead to an increase in the fatty acid concentration within the cell. This will stimulate esterification and the amount of triacylglycerol within the fat cell will increase. Such a process is mediated by a large number of

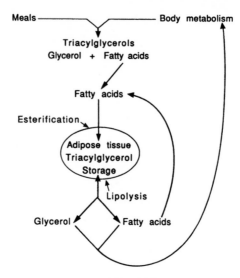

Figure 52

interactions, in which hormonal and nervous influences play a major role. In the case of increased energy requirement, fatty acids will be used in energy production. This will result in a decrease in the fatty acid concentration, which will stimulate the breakdown of triacylglycerols into glycerol and free fatty acids to compensate for this.

TRIACYLGLYCEROL METABOLISM

The process of binding fatty acids (esterification) in the form of triglyceride and their release from it is called the triglyceride/fatty acid cycle. The activity of this cycle is determined by the metabolic need for fatty acids for energy production and by the supply of fatty acids from external sources. The glycerol necessary for esterification is derived from glycolysis.

FATTY ACID METABOLISM

Free fatty acids are metabolized by aerobic metabolism within the citric acid cycle.

For this chain of metabolic steps, fatty acids are converted to fatty acyl CoA. This can enter the Krebs cycle where it is converted to acetyl coenzyme A. When fat oxidation is high there is increased production of acetyl CoA, which is converted into citrate—the first citric acid cycle intermediate. Acetyl CoA is known to inhibit the conversion of pyruvate to acetyl CoA. Additionally, citrate will inhibit glycolysis. Thus, increased

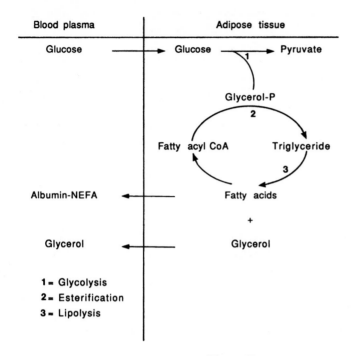

Figure 53

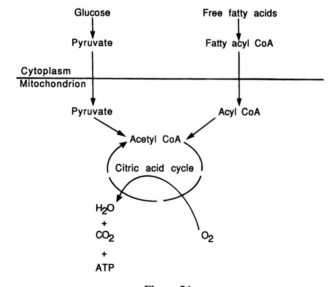

Figure 54

fatty acid oxidation inhibits both the rate of glycolysis and the first conversion step of pyruvate in the citric acid cycle. As a result, total carbohydrate oxidation will be reduced.

Conversely, increased carbohydrate metabolism, e.g. after intake of oral CHO, inhibits lipolysis, reduces the availability of fatty acids and thus their oxidation. In exercise metabolism these processes of carbohydrate and fat utilization are tightly coupled and controlled by nervous and hormonal mechanisms. They may be influenced by exogenous supply of either carbohydrate or fat, or by substances which stimulate the metabolism of either substrate.

PROTEIN

All protein in the body is *functional* protein. We do not have a protein store, as is the case with carbohydrate in the form of glycogen or fat stored as triacylglycerol in adipose tissue.

The amount of functional protein depends on organ function. Increased functioning, e.g. regular training stimuli for the heart or skeletal muscle, will result in a build-up of more contractile protein. As a result, the muscle will hypertrophy. Increased metabolic demand will lead to an increased number of enzymes and mitochondria, etc.

Amino acids are the building blocks of protein. The body cannot produce essential amino acids. Therefore, appropriate protein sources are required to supply these amino acids. Periods of enhanced growth are characterized by increased protein synthesis, periods of illness or inactivity are marked by increased protein degradation. In both instances the amount of amino acids and nitrogen needed is increased. An appropriate daily protein intake is therefore the key to the maintenance of nitrogen balance.

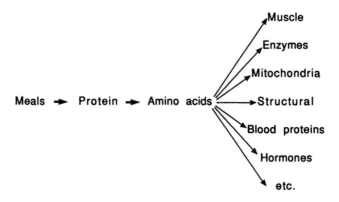

Figure 55

PROTEIN METABOLISM

The substances required for protein synthesis or resulting from protein breakdown are amino acids. Amino acids form a functional and metabolically available nitrogen pool in blood and in tissue fluids. Protein which is broken down, i.e. protein supplied with meals or protein within the body itself, results in a supply of amino acids into this pool. With an appropriate supply of amino acids, shortly after a meal, protein synthesis may be enhanced due to the combination of high insulin and appropriate amino acid supply. The amino acids that are not used in protein synthesis will either be oxidized or converted to carbohydrate and fat. A result of these processes is that the concentration of most amino acids in blood and tissue fluids is kept within a narrow range.

AMINO ACID OXIDATION

The general strategy of amino acid degradation is to produce metabolic intermediates that can be converted into glucose and fat and can be oxidized in the citric acid (Krebs) cycle. Most of the amino acids are oxidized within the liver and some of them—the branched chain amino acids—also in the muscle.

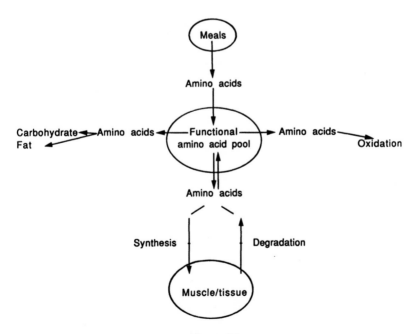

Figure 56

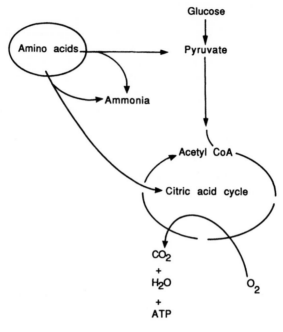

Figure 57

Amino acid oxidation takes place in the mitochondria and is always increased in periods of physical exercise. This oxidation will be increased when the carbohydrate availability for energetic processes becomes limited, as is the case with liver and glycogen depletion. The available evidence suggests that in such circumstances an increased amino acid requirement of 1.2–1.8 g/kg BW/day will be necessary to maintain nitrogen balance, for endurance athletes.

ENERGY METABOLISM

During the initial stage of sudden physical exercise, the extra amount of energy required is mainly produced by the breakdown of muscle glycogen to lactate. Blood glucose does not contribute substantially during the first minutes of exercise. The lactate formed is released into the bloodstream and taken up by the liver, the heart and by non-active muscle tissue, where it is either oxidized or resynthesized to glucose. At a later stage as glucose production from the liver is significant, muscle will increasingly use blood glucose for energy production. At this stage the glycogenolysis of the liver has to be increased.

Additionally, lipolysis in fat cells—initially a gradually increasing process—has led to high blood fatty acid levels, through which the contribution of fatty acids for energy production increases. Fatty acids become more and more oxidized in muscle and liver. Ketone bodies, which result from incomplete fat oxidation in the liver, are taken up from blood by the heart and the muscle for their final oxidation.

With increasing metabolic stress, especially in conditions of carbohydrate depletion, synthesis of protein may be decreased and the degradation of amino acids increases. Degradation of amino acids in muscle and liver

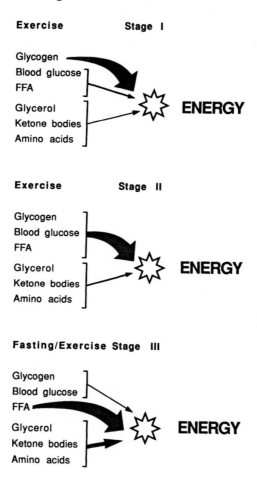

The size of the arrow indicates roughly the quantitative contibution to energy production

Figure 58

finally leads to the production of urea which will be excreted with urine and sweat. The carbon skeletons of the amino acids will enter the citric acid cycle in liver, where they will be used for gluconeogenesis, and muscle, where they will be oxidized.

With ongoing exercise and also during fasting, the endogenous carbohydrate stores in liver and muscle will become depleted. If no glucose were to be produced from gluconeogenic precursors in liver and kidney, the blood glucose level would drop sharply. Gluconeogenic precursors are amino acids, glycerol and lactate. At the same time, fat oxidation will be maximized, resulting in a reduced need for carbohydrate. Ketone bodies resulting from fat metabolism in the liver will be metabolized by heart, muscle and with prolonged fasting also in the brain. Under these circumstances, maximal work capacity will drop to approximately 50%, due to the lack of carbohydrate.

References

1 Ahlborg G., Felig P., Hagenfeldt L., *et al.* Substrate turnover during prolonged exercise in man. *J Clin Lab Invest* 1974; **53**: 1080–1090

2 Alhadeff L., Gualtieri T., Lipton M. Toxic effects of water-soluble vitamins. *Nut Revs* 1984; **2**: 33–40

3 Anderson R.A., Polansky M.M., Bryden N.A. *et al.* Strenuous exercise may increase dietary needs for chromium and zinc. In Katch F.I. (ed.). *Sport, Health and Nutrition.* 1984 Olympic Sci Cong Proc, Vol. 2. Champaign, Human Kinetics Publishers, 1986, pp 83–88

4 Anderson R.A., Polanski M.M., Bryden A. *et al.* Trace minerals and exercise: in Terjung R., Horton E.S. (eds). *Exercise, Nutrition and Energy Metabolism.* New York, Macmillan, 1988, pp 180–195

5 Anderson R.A. New insights on the trace elements, chromium, copper and zinc, and exercise. In: Brouns F., Saris W.H.M., Newsholme E.A. (eds). *Advances in Nutrition and Top Sport. Med Sport Sci* 1991; **32**: 38–58, Karger, Basel

6 Armstrong R.B. Muscle damage and endurance events. *Sports Med* 1986; **3**: 370–381

7 Auclair E., Sabatin P., Servan E., Guezennec C.Y. Metabolic effects of glucose, medium chain triglyceride and long chain triglyceride feeding before prolonged exercise in rats. *Eur J Appl Physiol* 1988; **57**: 126–131

8 Banister E.W., Cameron B.J. Exercise-induced hyperammonemia: peripheral and central effects. *Int J Sports Med* 1990; **11**(suppl 2): S129-S142

9 Barr S.I. Women, nutrition and exercise: a review of athletes' intakes and a discussion of energy balance in active women. *Prog Food Nutri Sci* 1987; **11**: 307–361

10 Beek van der E.J., Dokkum van W., Schrijver J. Marginal vitamin intake and physical performance in man. *Int J Sports Med* 1984; **5**: 28–31

11 Beek van der E.J. Vitamins and endurance training: food for running or faddish claims? *Sports Med* 1985; **2**: 175–197

12 Beek van der E.J., Dokkum van W., Schrijver J. Controlled vitamin C restriction and physical performance in volunteers. *J Am Coll Nutr* 1990; **9**(4): 332–339

13 Beek van der E.J. Vitamin supplementation and physical exercise performance. *J Sports Sci* 1991; **9**: 77–89

14 Bendich A. Exercise and free radicals: effects of antioxidant vitamins. In: *Advances in Nutrition and Top Sport.* Brouns F., Saris W.H.M., Newsholme E.A. (eds). *Med Sport Sci* 1991; **32**: 59–78, Karger, Basel

15 Bergström J., Hultman E. The effect of exercise on muscle glycogen and electrolytes in normals. *Scand J Clin Lab Invest* 1966; **18**: 16–20

16 Bergström J., Hultman E. A study of the glycogen metabolism during exercise in man. *Scand J Clin Lab Invest* 1967; **19**: 218–228

17 Bergström J., Hultman E. Synthesis of muscle glycogen in man after glucose and fructose infusion. *Acta Med Scand* 1967; **182**: 93–107

18 Bergström J., Fürst P., Holström B. *et al.* Influence of injury and nutrition on muscle water and electrolytes. *Ann Surg* 1981; **193**: 810–816

19　Bjorntorp P. Importance of fat as a support nutrient for energy: metabolism of athletes. *J Sports Sci* 1991; **9**: 71–76

20　Boje O. Doping: a study of the means employed to raise the level of performance in sport. *League of Nations Bulletin of the Health Organization* 1939; **8**: 439–469

21　Braun B., Clarkson P., Freedson P. *et al.* The effect of coenzyme Q10 supplementation on exercise performance, VO_2max, and lipid peroxidation in trained cyclists. *Int J Sport Nutr* 1991; **1**: 353–365.

22　Bremer J., Osmundsen H. Fatty acid oxidation and its regulation. In: Numa S. (ed.): *Fatty Acid Metabolism and its Regulation.* Amsterdam, Elsevier, 1984, pp 113–154

23　Brouns F., Saris W.H.M., ten Hoor F. Dietary problems in the case of strenuous exercise: Part 1 — A literature review, Part 2 — The athletes diet: nutrient dense enough? *J Sports Med Phys Fitness* 1986; **26**: 306–319

24　Brouns F., Saris W.H.M., Rehrer N.J. Abdominal complaints and gastro-intestinal function during long-lasting exercise. *Int J Sports Med* 1987; **8**: 175–189

25　Brouns F., Saris W.H.M., Beckers E. *et al.* Metabolic changes induced by sustained exhaustive cycling and diet manipulation. *Int J Sports Med* 1989; **10**(suppl 1): S49–S62

26　Brouns F., Rehrer N.J., Saris W.H.M. *et al.* Effect of carbohydrate intake during warming-up on the regulation of blood glucose during exercise. *Int J Sports Med* 1989; **10**(suppl 1): S68–S75

27　Brouns F., Rehrer N.J., Beckers E. *et al.* Reaktive Hypoglykämie. *Deutsche Zeitschr Sportmed* 1991; **42**(5): 188–200

28　Brouns F., Saris W.H.M. How vitamins affect performance. *J Sports Med Phys Fitness* 1989; **4**: 400–404

29　Brouns F., Beckers E., Wagenmakers A.J.M., Saris W.H.M. Ammonia accumulation during highly intensive long-lasting cycling: individual observations. *Int J Sports Med* 1990; **11** (suppl 2): S78–S84

30　Brouns F. Gastrointestinal symptoms in athletes: physiological and nutritional aspects. In: *Advances in Nutrition and Top Sport.* Brouns F., Saris W.H.M., Newsholme E.A. (eds). *Med Sport Sci* 1991; **32**: 166–199, Karger, Basel

31　Brouns F. Etiology of gastrointestinal disturbances during endurance events. *Scand J Med Sci Sports* 1991; **1**: 66–77

32　Brouns F. Heat — sweat — dehydration — rehydration: a praxis oriented approach. *J Sports Sci* 1991; **9**: 143–152

33　Bucci L., Hickson J., Pivarnik J. *et al.* Growth hormone release in bodybuilders after oral ornithine administration. *FASEB Journal* 1990; **4**: A397 (abstract)

34　Bucci L. Nutritional ergogenic aids. In: Hickson J., Wolinsky I. (eds). CRC Press, Boca Raton, FL, 1989

35　Campbell W.W., Anderson R.A. Effects of aerobic exercise and training on the trace minerals, chromium, zinc and copper. *Sports Med* 1987; **4**: 9–18

36　Chandler J., Hawkins J. The effect of bee pollen on physiological performance. *Int J Biosoc Med Res* 1984; **6**: 107–114

37　Clarkson P.M. Minerals: exercise performance and supplementation in athletes. *J Sports Sci* 1991; **9**: 91–116

38　Clement D.B., Sawchuk I.L. Iron status and sports performance. *Sports Med* 1984; **1**: 65–74

39　Costill D.L. Carbohydrates for exercise: dietary demands for optimal performance. *Int J Sports Med* 1988; **9**: 1–18

40   Costill D.L. Gastric emptying of fluids during exercise. In: Gisolfi C.V., Lamb
     D.R. (eds). *Perspectives in Exercise Science and Sports Medicine, Vol. 3, Fluid
     Homeostatis During Exercise.* Benchmark Press, Carmel, Indiana, 1990,
     pp 97–127

41   Couzy F., Lafargue P., Guezennec C.Y. Zinc metabolism in the athlete:
     influence of training, nutrition and other factors. *Int J Sports Med* 1990; **11:**
     263–266

42   Coyle E.F., Hamilton M. Fluid replacement during exercise: effects on
     physiological homeostasis and performance. In: Gisolfi C.V., Lamb D.R. (eds).
     *Perspectives in Exercise Science and Sports Medicine, Vol. 3, Fluid Homeostatis
     During Exercise.* Benchmark Press, Carmel, Indiana, 1990, pp 281–308

43   Coyle E.F., Coggan A.R. Effectiveness of carbohydrate feeding in delaying
     fatigue during prolonged exercise. *Sports Med* 1984; **1:** 446–458

44   Coyle E.F. Carbohydrate feedings: effects on metabolism, performance and
     recovery. In: *Advances in Nutrition and Top Sport.* Brouns F., Saris W.H.M.,
     Newsholme E.A. (eds). *Med Sport Sci* 1991; **32:** 1–14, Karger, Basel

45   Coyle E.F. Timing and method of increased carbohydrate intake to cope with
     heavy training, competition and recovery. *J Sports Sci* 1991; **9:** 29–52

46   Corley G., Demarest-Litchford M., Bazzarre T.L. Nutrition knowledge and
     dietary practices of college coaches. *Am Diet Assoc* 1990; **90:** 705–709

47   Crapo P.A. Simple versus complex carbohydrate use in the diabetic diet. *Ann
     Rev Nutr* 1985; **5:** 95–114

48   Décombaz J., Arnaud M.J., Milon H. *et al.* Energy metabolism of medium chain
     triglycerides versus carbohydrates during exercise. *Eur J Appl Physiol* 1983; **52:**
     9–14

49   Décombaz J., Sartori D., Arnaud M.J. *et al.* Oxidation and metabolic effects of
     fructose and glucose ingested before exercise. *Int J Sports Med* 1985; **6:** 282–286

50   Décombaz J., Gmuender B., Sierro G., Cerretelli P. Muscle carnitine after
     strenuous endurance exercise. *J Appl Physiol* 1992; **72(2):** 423–427

51   Demopoulous H., Santomier J., Seligman M., *et al.* Free radical pathology:
     Rationale and toxicology of antioxidants and other supplements in sports
     medicine and exercise science. In: Katch F. (ed.). *Sport Health and Nutrition,*
     Human Kinetics Publishers. Champaign, IL, 1986

52   Deutsche Gesellschaft für Ernährung, 1991. Empfehlungen für die Nährstoff-
     zufuhr, Umschau Verlag, Frankfurt

53   Dreher M.L., Dreher C.J., Berry J.W. Starch digestibility of foods: a nutritional
     perspective. *CRC Crit Rev* 1984; **20:** 47–71

54   Dunnet W., Crossen D. The bee pollen promise. *Runners World* 1980; **15(8):**
     53–54

55   Dwyer J. Nutritional status and alternative life-style diets with special
     reference to vegetarianism in the US. In: Rechcigl M. (ed.). *CRC Handbook of
     Nutritional Supplements, vol 1, Human Use.* CRC Press, New York, 1983

56   Eichner E.R. The anemias of athletes. *Physician Sports Med* 1986; **14(9):** 122–130

57   Eichner E.R. Other medical considerations in prolonged exercise. In: Lamb
     D.R., Murray R. (eds). *Perspectives in Exercise Science and Sports Medicine, Vol. 1,
     Prolonged Exercise.* Benchmark Press, Indianapolis, Indiana, 1988, pp 415–442

58   Erp-Baart van A.M.J., Saris W.H.M., Binkhorst R.A. *et al.* Nationwide survey
     on nutritional habits in elite athletes. Part I. Energy, carbohydrate, protein and
     fat intake. *Int J Sports Med* 1989; **10(suppl 1):** S3–S10

59   Erp-Baart van A.M.J., Saris W.H.M., Binkhorst R.A. *et al.* Nationwide survey
     on nutritional habits in elite athletes. Part II. Mineral and vitamin intake. *Int J
     Sports Med* 1989a; **10(suppl 1):** S11–S16

60   Erp-Baart van A.M.J. *Food Habits in Athletes.* Dissertation, University of Nijmegen, The Netherlands, 1992
61   Evans G.W. The effect of chromium picolinate on insulin controlled parameters in humans. *Int J Biosoc Med Res* 1989; **11**(2): 163–180
62   Faber M., Benadé A.J.S., van Eck M. Dietary intake, anthropometic measurements, and blood lipid values in weight training athletes (body builders). *Int J Sports Med* 1986; **7**(6): 342–346
63   Felig P., Wahren J. Fuel homeostasis in exercise. *New Engl J Med* 1975; **293**: 1078–1084
64   Felig P. Amino acid metabolism in exercise. *Ann NY Acad Sci* 1977; **301**: 56–63
65   Felig P. Hypoglycemia during prolonged exercise in normal man. *New Engl J Med* 1982; **306**: 895–910
66   Fern E.B., Bielinski R.N., Schutz Y. Effects of exaggerated amino acid and protein supply in man. *Experientia* 1991; **47**: 168–172
67   Galiano F. *et al.* Physiological, endocrine and performance effects of adding branched chain amino acids to a 6% carbohydrate-electrolyte beverage during prolonged cycling. *Med Sci Sports Exerc* 1991; **23**: S14 (abstract)
68   Gerster H. The role of vitamin C in athletic performance. *J Am Coll Nutr* 1989; **8**(6): 363–643
69   Gisolfi C.V., Summers R.W., Schedl H.P. Intestinal absorption of fluids during rest and exercise. In: Gisolfi C.V., Lamb D.R. (eds). *Perspectives in Exercise Science and Sports Medicine, Vol. 3, Fluid Homeostatis During Exercise.* Benchmark Press, Carmel, Indiana, 1990, pp 129–180
70   Gledhill N. Bicarbonate ingestion and anaerobic performance. *Sports Med* 1984; **1**: 177–180
71   Gollnick P.D. Energy metabolism and prolonged exercise. In: Lamb D.R., Murray R. (eds). *Perspectives in Exercise Science and Sports Medicine, Vol. 1, Prolonged Exercise.* Benchmark Press, Indianapolis, Indiana, 1988, pp 1–42
72   Grandjean A.C. The vegetarian athlete. *Physician Sports Med* 1987; **15**(5): 191–194
73   Greig C., Finch K., Jones D., *et al.* The effect of oral supplementation with L-carnitine on maximum and submaximum exercise capacity. *Eur J App Physiol* 1987; **56**: 457–460
74   Grossman S.B. (ed.). *Thirst and Sodium Appetite: Physiological Basis.* Academic Press, New York, 1990
75   Guezennec C.Y., Sabatin P., Duforez F. *et al.* Oxidation of corn starch, glucose and fructose ingested before exercise. *Med Sci Sports Exerc* 1989; **21**: 45–50
76   Guezennec C.Y., Sabatin P., Duforez F. *et al.* Role of starchy food type and structure on the metabolic response to physical exercise. *Med Sci Sports Exerc,* 1991
77   Guezennec C.Y., Léger C., Sabatin P. Lipid metabolism and performance. In: Atlan C., Béliveau L., Bouissou P. (eds). *Muscle Fatigue: Biochemical and Physiological Aspects.* Masson, Paris, 1991, pp 165–172
78   Hackman R.M., Keen C.L. Changes in serum zinc and copper levels after zinc supplementation in running and nonrunning men. In Katch F.I. (ed.). *Sport, Health and Nutrition.* 1984 Olympic Sci Cong Proc, Vol. 2. Human Kinetics Publishers, Champaign, IL, 1986, pp 89–100
79   Hargreaves M. Carbohydrates and exercise. *J Sport Sci* 1991; **9**: 18–28
80   Harper A.E., Zapalowski C. Metabolism of branched chain amino acids. In: Waterlow J.C., Stephen J.M.L. (eds). *Nitrogen Metabolism in Man.* Applied Science Publishers, London, 1981, pp 97–116

81   Hawkins C., *et al.* Oral arginine does not affect body composition or muscle function in male weight lifters. *Med Sci Sports Exerc* 1991; **23**: S15 (abstract)

82   Hawley J.A., Dennis S.C., Noakes T.D. Oxidation of carbohydrate ingested during prolonged endurance exercise. *Sports Med* 1992; **14**: 27–42

83   Haymes E.M. The use of vitamin and mineral supplements by athletes. *J Drugs Issues* 1980; **3**: 361–370

84   Haymes E.M. Vitamin and mineral supplementation to athletes. *Int J Sport Nutr* 1991; **1**: 146–169

85   Heigenhauser G., Jones N. Bicarbonate loading. In: Lamb D., Williams M. (eds). *Ergogenics: Enhancement of Performance in Exercise and Sport.* Brown & Benchmark, Dubuque, IA, 1991

86   Herbert V., Colman N., Jacob E. Folic acid and vitamin B12. In: Goodhart R., Shils M. (eds). *Modern Nutrition in Health and Disease.* Lea and Febiger, Philadelphia, 1980, pp 229–258

87   Heany R.P., Recher R.R., Saville P.D. Menopausal changes in calcium balance performance. *J Lab Clin Med* 1978; **92**: 953–963

88   Hubbard R.W., Szlyk P.C., Armstrong L.E. Influence of thirst and fluid palatability on fluid ingestion during exercise. In: Gisolfi C.V., Lamb D.R. (eds). *Perspectives in Exercise Science and Sports Medicine, Vol. 3, Fluid Homeostatis During Exercise.* Benchmark Press, Carmel, Indiana, 1990, pp 39–96

89   Hultman E. Studies on metabolism of glycogen and active phosphate in man with special reference to exercise and diet. *Scand J Clin Lab Invest* 1967; **19**(suppl. 94)

90   Hultman E. Dietary manipulations as an aid to preparation for competition. In: *Proc World Conference on Sportsmedicine,* Melbourne, 1974, pp 239–265

91   Hultman E. Liver glycogen in man: effect of different diets and muscular exercise. In: Pernow B., Saltin B. (eds). *Muscle Metabolism During Exercise.* Plenum, New York, 1981, pp 143–152

92   Ivy J.L., Costill D.L., Fink W. *et al.* Contribution of medium and long chain triglyceride intake to energy metabolism during prolonged exercise. *Int J Sports Med* 1980; **1**: 15–20

93   Jacob R.A., Harold H., Sandstead M.D. *et al.* Whole body surface loss of trace metals in normal males. *Am J Clin Nutr* 1981; **34**: 1379–1383

94   Jacobson B. Effect of amino acids on growth hormone release. *Physician Sports Med* 1990; **18**: 63–70

95   Janssen G.M.E., Scholte H.R., Vaandrager-Verduin M.H.M., Ross J.D. Muscle carnitine level in endurance training and running a marathon. *Int J Sports Med* 1989; **10**: S153–S155

96   Keen C.L., Hackman R.M. Trace elements in athletic performance. In Katch F.I. (ed.). *Sport, Health and Nutrition.* 1984 Olympic Sci Cong Proc, Vol. 2. Champaign, Human Kinetics Publishers, 1986, pp 51–66

97   Kieffer F. Trace elements: their importance for health and physical performance. *Deutsche Zschr für Sportmed* 1986; **37**: 118–123 (in German)

98   Kiens B., Raben A.B., Valeur A.K., Richter E. Benefit of dietary simple carbohydrates on the early post exercise glycogen repletion in male athletes. *Med Sci Sports Exerc* 1990; **22**: 588 (abstract)

99   Kreider R., *et al.* Effects of amino acid supplementation on substrate usage during ultraendurance triathlon performance. *Med Sci Sports Exerc* 1991; **23**: S16 (abstract)

100 Kreider R.B., Miller G.W., Williams M.H., Somma C.T., Nasser T. Effects of phosphate loading on oxygen uptake, ventilatory anaerobic threshold, and run performance. *Med Sci Sports Exerc* 1990; **22**: 250–255

101   Lamb D.R., Snyder, A.C. Muscle glycogen loading with a liquid carbohydrate supplement. *Int J Sports Nutr* 1991; 1: 52–60

102   Lampe J.W., Slavin J.L., Apple F.S. Elevated serum ferritin concentrations in master runners after a marathon race. *Int J Vitamin Nutr Res* 1986; 6: 395–398

103   Lane H.W. Some trace elements related to physical activity: zinc, copper, selenium, chromium and iodine. In: Hickson J.E, Wolinski I. (eds). *Nutrition in Exercise and Sport.* CRC Press, Florida, 1989, pp 301–307

104   Lau K. Phosphate disorders. In: Kokko J.P., Tannen R.L. (eds). *Fluids and Electrolytes.* WB Saunders, Philadelphia, 1986, pp 398–471

105   Lemon P.W.R. Nutrition for muscular development of young athletes. In: Gisolfi C.V., Lamb D.R. (eds). *Perspectives in Exercise Science and Sports Medicine, Vol. 2, Youth, Exercise, and Sport.* Benchmark Press, Indianapolis, Indiana, 1989, pp 369–400

106   Lemon P.W.R. Effect of exercise on protein requirements. *J Sports Sci* 1991; 9: 53–70

107   Lemon P.W.R. Does exercise alter dietary protein requirements? In: *Advances in Nutrition and Top Sport.* Brouns F., Saris W.H.M., Newsholme E.A. (eds). *Med Sport Sci* 1991; 32: 15–37, Karger, Basel

108   Lemon P.W.R. Protein and amino acid needs of the strength athlete. *Int J Sport Nutr* 1991; 1: 127–145

109   Levander O.A., Cheng L. Micronutrient interactions; vitamins, minerals and hazardous elements. *Ann NY Acad Sci* 1980; 355: 1–372

110   Mahalko J.R., Sandsted H.H., Johnson L.K., Milne D.B. Effect of a moderate increase in dietary protein on the retention and excretion of Ca, Cu, Fe, Mg, P and Zn by adult males. *Am J Clin Nutr* 1983; 37: 8–14

111   Massicotte D., Péronnet F., Brisson G., Hillaire-Marcel C. Exogenous 13 C lipids and 13 C glucose oxidized during prolonged exercise in man. *Med Sci Sports Exerc* 1990; 2: S52 (abstract)

112   Massicotte D., Péronnet F., Brisson G. *et al.* Oxidation of glucose polymer during exercise: comparison with glucose or fructose. *J Appl Physiol* 1989; 66: 179–183

113   Maughan R.J. Effects of diet composition on the performance of high intensity exercise. In: Monod H (ed.). *Nutrition et Sport.* Masson, Paris, 1990, pp 201–211

114   Maughan R.J. Fluid and electrolyte loss and replacement in exercise. *J Sports Sci* 1991; 9: 117–142

115   Maughan R.J. and Noakes T.D. Fluid replacement and exercise stress, a brief review of studies on fluid replacement and some guidelines for the athlete. *Sports Med* 1991a; 1: 16–31

116   Maughan R.J., Sadler D. The effects of oral administration of salts of aspartic acid on the metabolic response to prolonged exhausting exercise in man. *Int J Sports Med* 1983; 4: 119–123

117   Maughan R.J., Greenhaff P.L. High intensity exercise performance and acid-base balance: The influence of diet and induced metabolic alkalosis. In: Brouns F. (ed.). *Advances in Nutrition and Top Sport,* Med Sport Sci, Basel, Karger, 1991, vol 32, pp 147–165

117a Maughan R.J. Exercise-induced muscle cramp: a prospective biochemical study in marathon runners. *J Sports Sci* 1986; 4: 31–34

118   McDonald R., Keen C.L. Iron, zinc and magnesium nutrition and athletic performance. *Sports Med* 1988; 5: 171–184

119   McGilvery R.W. The use of fuels for muscular work. In: Howald H., Poortmans J.R. (eds). *Metabolic Adaptation to Prolonged Physical Exercise.* Proc. Second Int Symp on Biochemistry. Birkhäuser Verlag, Basel, 1973, pp 12–20

120 McNaughton L.R. Sodium citrate and anaerobic performance: implications of dosage. *Eur J Appl Physiol* 1990; **61**: 392–397

121 Medbo J.I., Serjested O. Plasma potassium changes with high intensity exercise. *J Physiol* 1990; **421**: 105–122

122 Mitchell M. *et al*. Effects of amino acid supplementation on metabolic responses to ultraendurance triathlon performance. *Med Sci Sports Exerc* 1991; **23**: S15 (abstract)

123 Moses F. The effect of exercise on the gastrointestinal tract. *Sports Med* 1990; **9**: 159–172

124 Mosora F., Lacroix M., Luyckx A.S. *et al*. Glucose oxidation in relation to the size of the oral glucose loading dose. *Metabolism* 1981; **30**: 1143–1149

125 Munro H.N. Metabolism and functions of amino acids in man—overview and synthesis. In: Blackburn G.L., Grant J.P., Vernon R.Y. (eds). *Amino Acids, Metabolism and Medical Applications*. John Wright, PSG Inc, Boston, 1983, pp 1–12

126 Murray R. The effects of consuming carbohydrate-electrolyte beverages on gastric emptying and fluid absorption during and following exercise. *Sports Med* 1987; **4**: 322–351

127 Murray R., Seifert J.G., Eddy D.E. *et al*. Carbohydrate feeding and exercise: effect of beverage carbohydrate content. *Eur J Appl Physiol* 1989; **59**: 152–158

128 Murray R., Paul L.P., Seifert J.G. *et al*. The effects of glucose, fructose and sucrose ingestion during exercise. *Med Sci Sports Exerc* 1989; **3**: 275–282

129 Nadel E.R. Temperature regulation and prolonged exercise. In: Lamb D.R., Murray R. (eds). *Perspectives in Exercise Science and Sports Medicine, Vol. 1, Prolonged Exercise*. Benchmark Press, Indianapolis, Indiana, 1988, pp 125–151

130 Nadel E.R., Mack G.W., Nose H. Influence of fluid replacement beverages on body fluid homeostasis during exercise and recovery. In: Gisolfi C.V., Lamb D.R. (eds). *Perspectives in Exercise Science and Sports Medicine, Vol. 3, Fluid Homeostatis During Exercise*. Benchmark Press, Carmel, Indiana, 1990, pp 181–205

131 National Research Council. *Recommended Dietary Allowances*, 10th edn. National Academy Press, Washington, 1989

132 Newhouse I.J., Clement D.B. Iron status in athletes. An update. *Sports Med* 1988; **5**: 337–352

133 Newsholme E.A., Start C. (eds). Regulation of glycogen metabolism. In: *Regulation in Metabolism*. John Wiley, Chichester, 1973, pp 146–194

134 Newsholme E.A., Start C. (eds). Adipose tissue and the regulation of fat metabolism. In: *Regulation in Metabolism*. John Wiley, Chichester, 1973a, pp 195–246

135 Newsholme E.A., Start C. (eds). Regulation of carbohydrate metabolism in liver. In: *Regulation in Metabolism*. John Wiley, Chichester, 1973b, pp 247–323

136 Newsholme E.A., Leech A.R. (eds). Integration of carbohydrate and lipid metabolism. In: *Biochemistry for the Medical Sciences*. John Wiley, Chichester, 1983, pp 336–356

137 Newsholme E.A., Leech A.R. (eds). Metabolism in exercise. In: *Biochemistry for the Medical Sciences*. John Wiley, Chichester, 1983, pp 357–381

138 Newsholme E.A., Leech A.R. (eds). The integration of metabolism during starvation, refeeding, and injury. In: *Biochemistry for the Medical Sciences*. John Wiley, Chichester, 1983, pp 536–561

139 Newsholme E.A., Parry-Billings M., McAndrew N. *et al*. A biochemical mechanism to explain some characteristics of overtraining. In: Brouns F., Saris W.H.M., Newsholme E.A. (eds). *Advances in Nutrition and Top Sport, Med Sport Sci* 1991; **32**: 79–93, Karger, Basel

140   Newsholme E. Effects of exercise on aspects of carbohydrate, fat, and amino acid metabolism. In: Bouchard C., Shephard R., Stephens T. *et al.* (eds). *Exercise, Fitness and Health.* Human Kinetics, Champaign, IL., 1990

141   Nieman D.C. Vegetarian dietary practices and endurance performance. *Am J Clin Nutr* 1988; **48**: 754–761

142   Noakes T.D., Goodwin N., Rayner B.L. *et al.* Water intoxication: a possible complication during endurance exercise. *Med Sci Sports Exerc* 1985; **17**: 370–375

143   Noakes T.D., Adams B.A., Myburgh K.H. *et al.* The danger of an inadequate water intake during prolonged exercise. A novel concept re-visited. *Eur J Appl Physiol* 1988; **57**: 210–219

144   Noakes T.D., Normann R.J., Buck R.H. *et al.* The incidence of hyponatremia during prolonged ultraendurance exercise. *Med Sci Sports Exerc* 1990; **22**: 165–170

145   Oppenheimer S., Hendrickse R. The clinical effects of iron deficiency and iron supplementation. *Nutr Abs Rev* 1983; **53**: 585–598, series A

146   Otto R., Shores K., Wygard J. *et al.* The effects of L-carnitine supplementation on endurance exercise. *Med Sci Sports Exerc* 1987; **19**: S87 (abstract)

147   Oyono-Enguelle S., Freund H., Ott C. *et al.* Prolonged submaximal exercise and L-carnitine in humans. *Eur J Appl Physiol* 1988; **58**: 53–61

148   Pallikarakis N., Jandrain B., Pirnay F. *et al.* Remarkable metabolic availability of oral glucose during long duration exercise in humans. *J Appl Physiol* 1986; **60**: 1035–1042

149   Parr R.B., Porter M.A., Hodgson S.C. Nutrition knowledge and practice of coaches, trainers and athletes. *Phys Sportsmed* 1984; **12**: 127–138

150   Pate R.R. Sports anemia: a review of the current research literature. *Phys Sportsmed* 1983; **11**: 115–131

151   Pate R.R., Sargent R.G., Baldwin C., Burgess M.L. Dietary intake of women runners. *Int J Sports Med* 1990; **11**(6): 461–466

152   Puhl J.L., Handel van P.J., Williams L.L. *et al.* Iron status and training. In: Butts N.K., Gushiken T.T., Zarins B. (eds). *The Elite Athlete.* Champaign, IL, Life Enhancement Publications, 1985, pp 209–238

153   Rehrer N.J., Beckers E., Brouns F. *et al.* Exercise and training effects on gastric emptying of carbohydrate beverages. *Med Sci Sports Exerc* 1989; **21**: 540–549

154   Rehrer N.J., Beckers E.J., Brouns F., Saris W.H.M., ten Hoor F. Effects of electrolytes in carbohydrate beverages on gastric emptying and secretion. *Med Sci Sports Exerc* 1993; **25**(1): 42–51

155   Rehrer N.J. Aspects of dehydration and rehydration during exercise. In: *Advances in Nutrition and Top Sport.* Brouns F., Saris W.H.M., Newsholme E.A. (eds). *Med Sport Sci* 1991; **32**: 128–146, Karger, Basel

156   Rehrer N.J., van Kemenade M.C., Meester T.A. *et al.* Gastrointestinal complaints in relation to dietary intakes in triathletes. *Int J Sport Nutr* 1992; **2**: 48–59

157   Rennie M.J., Edwards R.H.T., Halliday D. *et al.* Protein metabolism during exercise. In: Waterlow J.C., Stephen J.M.L. (eds). *Nitrogen Metabolism in Man.* Applied Science Publishers, London, 1981, pp 509–524

158   Richter E.A., Sonne B., Plough T. *et al.* Regulation of carbohydrate metabolism in exercise. In: Saltin B. (ed.). *Biochemistry of Exercise*, Vol. 6. Champaign IL, Human Kinetics Publishers, 1986, pp 151–166

159   Roberts J. The effect of coenzyme Q10 on exercise performance. *Med Sci Sports Exerc* 1990; **22**: S87 (abstract)

160 Rumessen J.J., Gudmand-Hoyer E. Absorption capacity of fructose in healthy adults. Comparison with sucrose and its constituent monosaccharides. *Gut* 1986; **27**: 1161–1168

161 Russel McR. D., Jeejeeboy K.N. The assessment of the functional consequences of malnutrition. *Nutr Abstr and Rev in Clin Nutr*, series A, 1983; **53**(10): 863–877

162 Sabatin P., Portero P., Gilles D. *et al.* Metabolic and hormonal responses to lipid and carbohydrate diets during exercise in man. *Med Sci Sports Exerc* 1987; **19**(3): 218–223

163 Saris W.H.M., Brouns F. Nutritional concerns for the young athlete. In: Rutenfranz J., Mocellin R., Klinit F. (eds). *International Series on Sports Sciences. Vol 17, Children and Exercise XII.* Human Kinetics, Champaign, IL, 1986, pp 11–18

164 Saris W.H.M., Schrijver J., Erp Baart van M.A., Brouns F. (1989) Adequacy of vitamin supply under maximal sustained workloads: the Tour de France. In: Walter P., Brubacher, G.B., Stähelin H.B. (eds). *Elevated Dosages of Vitamins.* Huber Publishers, Toronto

165 Saris W.H.M, van Erp Baart M.A., Brouns F. *et al.* Study on food intake and energy expenditure during extreme sustained exercise: the Tour de France. *Int J Sports Med* 1989; **10**(suppl), S26–S31

166 Sawka M.N. Body fluid responses and hypohydration during exercise-heat stress. In: Pandolf K.B., Sawka M.N., Gonzalez R.R. (eds). *Human Performance Physiology and Environmental Medicine at Terrestrial Extremes.* Benchmark Press, Indianapolis, Indiana, 1988, pp 227–266

167 Sawka M.N., Wenger C.B. Physiological responses to acute exercise-heat stress. In: Pandolf K.B., Sawka M.N., Gonzalez R.R. (eds). *Human Performance Physiology and Environmental Medicine at Terrestrial Extremes.* Benchmark Press, Indianapolis, Indiana, 1988a, pp 97–151

168 Sawka M.N., Pandolf K.B. Effects of body water loss on physiological function and exercise performance. In: Gisolfi C.V., Lamb D.R. (eds). *Perspectives in Exercise Science and Sports Medicine, Vol 3, Fluid Homeostatis During Exercise.* Benchmark Press, Carmel, Indiana, 1990, pp 1–38

169 Schoutens A., Laurent E., Poortmans J.R. Effect of inactivity and exercise on bone. *Sports Med* 1989; **7**: 71–81

170 Segura R., Ventura J. Effect of L-tryptophan supplementation on exercise performance. *Int J Sports Med* 1988; **9**: 301–305

171 Shephard R.J. Vitamin E and athletic performance. *J Sports Med* 1983; **23**: 461–470

172 Sherman W.M. Carbohydrates, muscle glycogen, and muscle glycogen supercompensation. In: Williams M.H. (ed.). *Ergogenic Aids in Sport.* Human Kinetics Publishers, Champaign, IL, 1983, pp 3–26

173 Sherman W.M., Lamb D.R. Nutrition and prolonged exercise. In: Lamb D.R., Murray R. (eds). *Perspectives in Exercise Science and Sports Medicine, Vol. 1, Prolonged Exercise.* Benchmark Press, Indianapolis, Indiana, 1988, pp 213–280

174 Sherman W.M. and Wimer G.S. Insufficient dietary carbohydrate during training: does it impair performance? *Int J Sports Nutr* 1991; **1**: 28–44

175 Shores K., Otto R., Wygard J. *et al.* Effect of L-carnitine supplementation on maximal oxygen consumption and free fatty acid serum levels. *Med Sci Sports Exerc* 1987; **19**: S60 (abstract)

176 Short S.H., Short W.R. Four-year study of university athletes' dietary intake. *Am Diet Assoc* 1983; **82**(6): 632–645

177 Smith J.C., Morris E.R, Ellis R. Zinc requirements, bioavailabilies and recommended dietary allowances. In: Prasad A.A. (ed.). *Zinc Deficiency in Human Subjects.* Alan R Liss, New York, 1983, pp 147–169

178   Spencer H., Kramer L., Perakis E. *et al.* Plasma levels of zinc during starvation. *Fed Proc* 1982; **41**: 347

179   Steben R., Wells J., Harless I. The effects of bee pollen tablets on the improvement of certain blood factors and performance of male collegiate swimmers. *J Natl Athletic Trainers Assoc* 1976; **11**: 124–126

180   Storlie J. Nutrition assessment of athletes: a model for integrating nutrition and physical performance indicators. *Int J Sport Nutr* 1991; **1**: 192–204

181   Sutton J.R. Clinical implications of fluid imbalance. In: Gisolfi C.V., Lamb D.R. (eds): *Perspectives in Exercise Science and Sports Medicine, Vol. 3, Fluid Homeostatis During Exercise*, Benchmark Press, Carmel, Indiana, 1990, pp 425–455

182   Tarnopolsky M.A., MacDougall J.D., Atkinson S.A. Influence of protein intake and training status on nitrogen balance and lean body mass. *J Appl Physiol* 1988; **64**(1): 187–193

183   Trichopoulou A., Vassilakos T. Recommended dietary intakes in the European Community member states. *Eur J Clin Nutr* 1990; suppl. 2: 51–101

184   Vandewalle L. *et al.* Effect of branched-chain amino acid supplements on exercise performance in glycogen depleted subjects. *Med Sci Sports Exerc* 1991; **23**: S116 (abstract)

185   Vollestad N.K, Serjested O.M. Plasma K+ shifts in muscle and blood during and after exercise. *Int J Sports Med* 1989; **10**(suppl 2): 101

186   Wade C.E., Freund B.J. Hormonal control of blood volume during and following exercise. In: Gisolfi C.V., Lamb D.R. (eds). *Perspectives in Exercise Science and Sports Medicine, Vol. 3, Fluid Homeostatis During Exercise*, Benchmark Press, Carmel, Indiana, 1990, pp 207–245

187   Wagenmakers A.J.M., Coakley J.H., Edwards R.H.T. Metabolism of branched-chain amino acids and ammonia during exercise: clues from McArdle's disease. *Int J Sports Med* 1990; **11**(suppl 2): S101–S113

188   Wagenmakers A.J.M., Beckers E.J., Brouns F. *et al.* Carbohydrate supplementation, glycogen depletion, and amino acid metabolism during exercise. *Am J Physiol* 1991; **260**: E883–E890

189   Wagenmakers A.J.M. A role of amino acids and ammonia in mechanisms of fatigue. In: Marconnet P., Saltin B., Komi P. (eds). *Local Fatigue in Exercise and Training*, 4th Int Symposium on Exercise and Sport Biology. *Med Sport Sci*, 1991; **34**.

190   Wagenmakers A.J.M. L-Carnitine supplementation and performance in man. In: Brouns F. (ed.). *Advances in Nutrition and Top Sport, Vol. 32*. Karger, Basel, 1991, pp 110–127

191   Walberg J.L., Leidy M.K., Sturgill D.J., *et al.* Macronutrient content of a hypoenergy diet affects nitrogen retention and muscle function in weight lifters. *Int J Sports Med* 1988; **9**(4): 261–266

192   Warren B., *et al.* The effect of amino acid supplementation on physiological responses of elite junior weightlifters. *Med Sci Sports Exerc* 1991; **23**: S15 (abstract)

193   Wassermann D.H., Geer R.J., Williams P.E. *et al.* Interaction of gut and liver in nitrogen metabolism during exercise. *Metabolism* 1991; **40**: 307–314

194   Waterlow J.C. Metabolic adaptation to low intakes of energy and protein. *Ann Rev Nutr* 1986; **6**: 495–526

195   Weight L.M., Myburgh K.M., Noakes T.D. Vitamin and mineral supplementation: effect on the running performance of trained athletes. *Am J Clin Nutr* 1988; **47**: 192–195

196   Wenger C.B. Human heat acclimatization. In: Pandolf K.B., Sawka M.N., Gonzalez R.R. (eds). *Human Performance Physiology and Environmental Medicine*

at Terrestrial Extremes. Benchmark Press, Indianapolis, Indiana, 1988, pp 153–197

197 Wesson M., McNaughton L., Davies P. *et al.* Effects of oral administration of aspartic acid salts on the endurance capacity of trained athletes. *Res Quarterly Exerc Sport* 1988; 59: 234–239

198 White S.L., Maloney S.K. Promoting healthy diets and active lives to hard-to-reach groups: market research study. *Public Health Reports* 1990; 105(3): 224–231

199 Williams M.H. *Nutritional Aspects of Human Physical and Athletic Performance,* 2nd edn. Charles C. Thomas, Springfield, IL, 1985

200 Williams M., Kreider R., Hunter D., *et al.* Effect of oral inosine supplementation on 3-mile treadmill run performance and VO_2 peak. *Med Sci Sports Exerc* 1990; 22: 517–522

201 Williams M. Ergogenic aids. In: Berning J., Steen S. (eds). *Sports Nutrition for the 90s: The Health Professional's Handbook.* Aspen Publishers, Gaithersburg, MD, 1991

202 Williams M.H. *Nutrition for Fitness and Sport,* 3rd edn. Wm. C. Brown Publishers, Dubuque, Iowa, 1992

203 Wilmore J., Freund B.J. Nutritional enhancement of physical performance. *Nutr Abs Rev Clin Nutr* 1984 series A; 54(1): 1–16

204 Wolfe R.R., Richard D., Goodenough D. *et al.* Protein dynamics in stress. In: Blackburn G.L., Grant J.P., Vernon R.Y. (eds). *Amino Acids, Metabolism and Medical Applications.* John Wright, PSG Inc, Boston, 1983, pp 396–413

205 Woodhouse M., Williams M., Jackson C. The effects of varying doses of orally ingested bee pollen extract upon selected performance variables. *Athletic Training* 1987; 22: 26–28

206 Wretlind A. Nutrition problems in healthy adults with low activity and low caloric consumption. In: Blix G. (ed.). *Nutrition and Physical Activity.* Almquist and Wiksels, 1976, pp 114–131

207 Wurtman R.J., Lewis M.C. Exercise, plasma composition and neurotransmission. In: Brouns F., Newsholme E.A., Saris W.H.M. (eds). *Advances in Nutrition and Topsport. Med Sport Sci,* Vol. 32. Karger, Basel, 1991, pp 94–109

208 Wyss V., *et al.* Effects of L-carnitine administration on VO_2max and the aerobic–anaerobic threshold in normoxia and acute hypoxia. *Eur J Appl Phys* 1990; 60: 1–6

209 Yoshimura H., Inoue T., Yamada T., Shiraki K. Anemia during hard physical training (sports anemia) and its causal mechanisms with special reference to protein nutrition. *World Rev Nutr Diet* 1980; 35: 1–86

210 Zuliani U., Bonetti A., Campana M. *et al.* The influence of ubiquinone (CoQ10) on the metabolic response to work. *J Sports Med Phys Fitness* 1989; 29: 57–61

211 Maughan R.J. Exercise induced muscle cramp: a prospective biochemical study in marathon runners. *J Sports Sci* 1986; 4: 31–34

212 Brouns F., Saris W.H.M., Schneider H. Rationale for upper limits of electrolyte replacement during exercise. *Int J Sports Nutr* 1992; 2: 229–238.

213 Wagenmakers A.J.M., Brouns F., Saris W.H.M *et al.* Oxidation rates of oral carbohydrates during prolonged exercise in men. *J Appl Physiol* 1993; 75(6): 2774–2780

214 Brouns F., Beckers E., Knopfli B., Villiger B. Rehydration during exercise: effect of electrolyte supplementation on selective blood parameters. *Med Sci Sports Exerc* 23(2), suppl 584, 1991

215 Rehrer N., Brouns F., Beckers E.J., *et al.* The influence of beverage composition and gastrointestinal function on fluid and nutrient availability during exercise. *Scand J Sports Med* 1994; 4: 159–172

216 Kazunori N., Clarckson P.M. Changes in plasma zinc following high force eccentric exercise. *Natl J Sport Nutr* 1992; 2: 175–184

217 Chryssanthopoulos C.C., Williams W., Wilson *et al.* Comparison between carbohydrate feedings before and during exercise on running performance during a 30 km treadmill time trial. *Int J Sport Nutr* 1994; 4: 374–386

218 Yaspelkis B.B., Ivy J.L. Effect of carbohydrate supplements and water on exercising metabolism in the heat. *J Appl Physiol* 1991; 71: 680–687

219 Leijssen D.P.C., Saris W.H.M., Jeukendrup A.E., Wagenmakers A.J.M. Oxidation of exogenous [13C]galactose and [13C]glucose during exercise. *J Appl Physiol* 1995; 79: 720–725

220 Brouns F., Van der Vusse G. Review article: Utilization of lipids during exercise in human subjects: metabolic and dietary constraints. *Brit J Nutr* 1998; 79: 117–128

221 Hawley J.A., Brouns F., Jeukendrup A. Strategies to enhance fat utilization during exercise. *Sports Med* 1998; 25(4): 241–257

222 Jeukendrup A., Wagenmakers A.J.M., Saris W.H.M. Fat metabolism during exercise: a review. Part 1, *Int J Sports Med* 1998; 19: 231–244, Part 2, *Int J Sports Med* 1998; 19: 1998, 293–302, Part 3, *Int J Sports Med* 1998; 19: 371–379

223 Oostenbrug G., Mensink R.P., Hardeman M.R. *et al.* Exercise performance, red blood cell deformability and lipid peroxidation: effects of fish oil and vitamin E. *J Appl Physiol* 1997; 83(3): 746–752.

224 Lemon, P.W.R. Do athletes need more dietary protein and amino acids? *Int J Sport Nutr* 5: S39–S61

225 Lemon P.W.R. Protein metabolism during exercise. In: Garret W.W., Kirkendall D.T. (eds). *Exercise and Sport Science*. Lippincott Williams and Wilkins Publishers, Philadelphia, 2000, pp 19–27

226 Noakes T.D. Hyponatremia during endurance running: a physiological and clinical interpretation. *Med Sci Sports Exerc* 1993; 24: 403–405

227 Brouns F., Kovacs E. Functional drinks for athletes, *Trends Food Sci Technol* 1997; 8: 414–421

228 Terblanche S., Moakes T., Dennis S.C. *et al.* Failure of magnesium supplementation to influence marathon running performance as recovery in magnesium replete subjects. *Int J Sport Nutr* 1992; 2: 154–164

229 Nosaka K., Clarkson P.M. Changes in plasma zinc following high force excentric exercise. *Int J Sport Nutr* 1992; 2: 175–184

230 Tipton K., Green N.R., Haynes E.M., Waller M. Zinc losses in sweat of athletes. Exercising in hot and neutral temperatures. *Int J Sport Nutr* 1993; 3: 261–271

231 Dressendorfer R.H., Wade C.E., Keen C.L., Scaff H.J. Plasma mineral levels in marathon runners during a 20-day road race. *Phys. Sportsmed* 1982; 10(6): 113–118

232 Couzy F., Lafargue P., Guezennec C.Y. Zinc metabolism in the athlete: influence of training, nutrition and other factors. *Int J Sport Nutr* 1990; 11: 263–266

233 Ohno H., Sato Y., Ishikawa M. *et al.* Training effects of blood zinc levels in humans. *J Sportsmed Phys Fit* 1990; 30: 247–253

234 Dressendorfer R.H. Sockolov R. Hypozincemia in runners. *Phys Sportsmed* 1980; 8(4): 97–100

235 Haralambie G. Serum zinc in athletes in training. *Int J Sports Med* 1981; 2: 135–138

236 Peters E.M., Goetsche J.M., Grobbelaar B, Noakes T.D. Vitamin C supplementation reduces the incidence of post race symptoms of upper respiratory tract infection in ultramarathon runners. *Am J Clin Nutr* 1993; 57(2): 170–174

237 Brouns F. Functional foods for athletes. *Trends Food Sci Technol* 1997; **8**: 358–363

238 Craciun A.M., Knapen M.H.J., Brouns F., Vermeer C. Improved bone metabolism in female athletes after vitamin K supplementation. *Int J Sports Med* 1998; **19**: 479–484

239 Gross M., Kormann B., Zöllner N. Ribose administration during exercise: effects on substrates and products of energy metabolism in healthy subjects and a patient with myoadenylate deaminase deficiency. *Klin Wochenschr* 1991; **69**: 151–155

240 Gross M., Reiter S., Zöllner N. Metabolism of D-ribose administered continuously to healthy persons and to patients with myoadenylate deaminase deficiency. *Klin Wochenschr* 1989; **67**: 1205–1213

241 Hellsten-Westling Y., Norman B., Balsom P.D., Sjödin B. Decreased resting levels of adenine nucleotides in human skeletal muscle after high-intensity training. *J Appl Physiol* 1993; **74**: 2523–2528

242 Hellsten Y., Richter E.A., Kiens B., Bangsbo J. AMP deamination and purine exchange in human skeletal muscle during and after intense exercise. *J Physiol. (London)* 1999; **520**: 909–920

243 Ibel H., Zimmer H.-G. Metabolic recovery following temporary regional myocardial ischemia in the rat. *J Mol Cell Cardiol* 1986; **18**: 61–65

244 Mauser M., Hoffmeister H.M., Nienaber C., Schaper W. Influence of ribose, adenosine, and 'AICAR' on the rate of myocardial adenosine triphosphate synthesis during reperfusion after coronary artery occlusion in the dog. *Circ Res* 1985; **56**: 220–230

245 Pasque M.K., Spray T.L., Pellom G.L. et al. Ribose-enhanced myocardial recovery following ischemia in the isolated working rat heart. *J Thorac Cardiovasc Surg* 1982; **83**: 398

246 Segal S., Foley J. The metabolism of D-ribose in man. *J Clin Invest* 1958; **37**: 719–735

247 St Cyr J.A., Bianco R.W., Schneider J.R. et al. Enhanced high energy phosphate recovery with ribose infusion after global myocardial ischemia in a canine model. *J Surg Res* 1989; **46**: 157–162

248 Tan Z.T., Wang, X.W. Verapamil, ribose and adenine enhance resynthesis of postischemic myocardial ATP. *Life Sci* 1994; **55**: 345–349

249 Tullson P.C., Bangsbo J., Hellsten Y., Richter, E.A. IMP metabolism in human skeletal muscle after exhaustive exercise. *J Appl Physiol* 1995; **78**: 146–152

250 Tullson P.C., Terjung R.L. Adenine nucleotide synthesis in exercising and endurance-trained skeletal muscle. *Am J Physiol (Cell Physiol)* 1991; **261**: C342-C347

251 Ward H.B., St Cyr J.A., Cogordan J.A. et al. Recovery of adenine nucleotide levels after global myocardial ischemia in dogs. *Surgery* 1984; **96**: 248–255

252 Zimmer H.-G. Significance of the 5-phosphoribosyl-1-pyrophosphate pool for cardiac purine and pyrimidine nucleotide synthesis: studies with ribose, adenine, inosine, and orotic acid in rats. *Cardiovasc Drugs Ther* 1998; **12**: 179–187

253 Zimmer H.-G., Gerlach E. Stimulation of myocardial adenine nucleotide biosynthesis by pentoses and pentitols. *Pflügers Arch* 1978; **376**: 223–227

254 Zimmer H.-G., Martius P.A., Marschner G. Myocardial infarction in rats: effects of metabolic and pharmacologic interventions. *Basic Res Cardiol* 1989; **84**: 332–343

255 Op 't Eijnde B., van Leemputte M., Brouns F. et al. No effects of oral ribose supplementation on repeated maximal exercise and *de novo* ATP resynthesis. *J Appl Physiol* 2001; **91**: 2275–2281

256 Zimmer H.G., Ibel H. Ribose accelerates the repletion of the ATP pool during recovery from reversible ischemia of the rat myocardium. *J Mol Cel Cardiol* 1984; **16**: 863–866

257 Pliml W. *et al.* Ribose improves work tolerance in cardiac patients. *Lancet* 1992; 507–510

258 St Cyr J. *et al.* Ribose metabolism. *Adv Exp Med Biol* 1986; **194**: 471–481

259 Bangbo J., Gollnick P.D., Graham T.E. *et al.* Anaerobic energy production and O_2 deficit relationship during exhaustive exercise in humans. *J Physiol* 1990; **42**: 539–559

260 Bogdanis G.C., Nevill M.E., Boobis L.H., Lakomy H.K. Contribution of phosphocreatine and aerobic metabolism to energy supply during repeated sprint exercise. *J Appl Physiol* 1996; **80**(3): 876–884

261 Balsom P.D., Soderlund B., Ekblom B. Creatine in humans, with special reference to creatine supplementation. *Sports Med* 1994; **18**: 268–280

262 Greenhaff P.L. Creatine and its application as an ergogenic aid. *Int J Sports Nutr* 1995; 5(suppl): S100–S110

263 Maughan R. Nutritional ergogenic aids and exercise performance. *Nut Res Rev* 1999; 12(2): 255–276

264 Wagenmakers A.J.M. Nutritional supplements: effects on exercise performance and metabolism. In Lamb D.R. (ed.). *Perspectives in Exercise Science and Performance*, Vol. 12. Cooper Publishing Group, 1999, pp 207–259

265 Williams M.H., Kreider R.B., Branch J.D. *Creatine the Power Supplement.* Human Kinetics Publishing, Campaign, IL, 1999

266 Paoletti R., Poli A., Jackson A.S. *Creatine: From Basic Science to Clinical Application.* Kluwer Academic Publishers, Dordrecht, Netherlands, 2000

267 Maughan R. Creatine supplementation and exercise performance. *Int J Sports Nutr* 1995; **5**: 94–101

268 Mujika I., Padilla S. Creatine supplementation as an ergogenic aid for sports performance in highly trained athletes, a critical review. *Int J Sports Med* 1997; **18**: 491–496

269 Poortmans J. Long term oral creatine supplementation does not impair renal function in athletes. Letter to the editor: reply. *Med Sci Sports Exerc* 1999; 248–249

270 Nieuwenhoven van M.A., Brouns F., Brummer R.J. Exercise and gastro-intestinal function. In: Garret W.E., Kirkendall D.T. (eds). *Exercise and Sport Science.* Lippincott Williams and Wilkins Publishers, Philadelphia, 2000, pp 191–216.

271 Kanter M. Free radicals, exercise and antioxidant supplementation. *Int J Sport Nutr* 1994; **4**: 205–220.

272 Jenkins R.R. Free radical chemistry: relationship to exercise. *Sports Med* 1988; **5**: 156–170

273 Halliwell B., Gutteridge J.M.C. *Free Radicals in Biology and Medicine.* Clarenden Press, Oxford, 1985

274 Sen C.K., Packer L., Hanninen O. *Exercise and Oxygen Toxicity.* Elsevier, Amsterdam, 1994

275 Karlsson J. *Antioxidants Exercise.* Human Kinetics Publishing, Champaign IL, 1997

276 Sen C.K., Sashwati R., Packer L. Exercise induced oxidative stress and antioxidant nutrients. In: *Nutrition in Sport: Encyclopaedia of Sports Medicine.* An IOC medical commission publication. Blackwell Science, Oxford

277 Li Li Ji. Free radicals and antioxidants in exercise and sports. In Garret W.E., Kirkendall D.T. (eds). *Exercise and Sport Science.* Lippincott Williams and Wilkins Publishers, Philadelphia, 2000, pp 299–317

278 Muggli R. (ed.). Free radical tissue damage: the protective role of antioxidant nutrients. *Free Radicals and Antioxidants in Nutrition.* Richelieu Press, London, 1993, pp 189–204

279 National Research Council, committee on diet and health. National Academy Press, Washington DC, 1989

280 Conning D. Antioxidant nutrients in health and disease. BNF Briefing Paper 1991; **25**: 2–15

281 Bendich A. Carotenoids and the immune response. *J Nutr* 1989; **119**: 112–115

282 Bendich A. Symposium conclusions: biological actions of carotenoids. *J Nutr* 1989; **119**: 135–136

283 Brady P.S., Brady L.J., Ullrey D.E. Selenium, vitamin E and the response to swimming stress in the rat. *J Nutr* 1979; **109**: 1103–1109

284 Burton G.W. Antioxidant action of carotenoids. *J Nutr* 1989; **119**: 109–111

285 Dillard C.J., Litov R.E. Savin W.M. *et al.* Effects of exercise, vitamin E and ozone on pulmonary function and lipid peroxidation. *J Appl Physiol: Respirat Environ Exerc Physiol* 1978; **45**: 927–932

286 Fellman J.H., Roth E.S. The biological oxidation of hypotaurine to taurine. *Taurine: Biological Actions and Clinical Perspectives.* 1985, pp 71–82

287 Frei B., England L., Ames B.N. Ascorbate is an outstanding antioxidant in human blood plasma. *Proc Natl Acad Sci* 1989; **86**: 6377–6381

288 Jackson M.J. Muscle damage during exercise: possible role of free radicals and protective effect of vitamin E. *Proc Nutr Soc* 1987; **46**: 77–80

289 Jenkins R.R. The role of superoxide dismutase and catalase in muscle fatigue. *Biochemistry of Exercise.* Human Kinetics Publishers, Champaign, 1983, pp 467–471

290 Jenkins R.R., Krause K., Schofield L.S. Influence of exercise on clearance of oxidant stress products and loosely bound iron. *Med Sci Sports Exerc.* 1993; **25**: 213–217.

291 Johnson M.A., Fischer J.A., Kays S.E. Is copper an antioxidant nutrient? *Crit Rev Food Sci Nutr* 1992; **32**: 1–31.

292 Lang J.K., Gohil K., Packer L., Burk R.F. Selenium deficiency, endurance exercise capacity, and antioxidant status in rats. *Am Physiol Soc* 1987; **63**: 2532–2535

293 Lemoyne M., Gossum van A., Kurian R. *et al.* Abstracts on vitamin metabolism. Breath pentane analysis as an index of lipid peroxidation: a functional test of vitamin E status. *Am J Clin Nutr* 1987; **46**: 257–272

294 Novelli G.P., Bracciotti G., Falsini S. Spin-trappers and vitamin E prolong endurance to muscle fatigue in mice. *Free Radical Biol Med* 1990; **8**: 9–13

295 Olson J.A. Biological actions of carotenoids. *J Nutr* 1989; **119**: 94–95.

296 Packer L. Vitamin E, physical exercise and tissue damage in animals. *Med Biol* 1984; **62**: 105–109

297 Packer L. Protective role of vitamin E in biological systems. *Am J Clin Nutr* 1991; **53**: 1050S–1055S

298 Parker R.S. Carotenoids in human blood and tissues. *J Nutr* 1989; **119**: 101–104

299 Robertson J.D., Maughan R.F., Duthie G.G., Morrice P.C. Increased blood antioxidant systems of runners in response to training load. *Clin Sci* 1991; **80**: 611–618

300 Shahidi F., Wanasundara P.K.J.D. Phenolic antioxidants. *Crit Rev Food Sci Nutr* 1992; **32**: 67–103

301 Bargossi A.M., Fiorella P.L., Grossi G, *et al.* Antioxidant effects of exogenous ubiquinone (Q10) in high level endurance runners. In *Free Radicals and Antioxidants in Nutrition.* Richelieu Press, London, 1993, pp 63–74

302   Karlsson J. *Coenzyme Q*. Vitaminex AB, Norrkoping, Sweden, 1993
303   Powers S.K., Leeuwenburg van C. Exercise induced alterations in skeletal muscle antioxidant capacity: a brief review. *Med Sci Sports Exerc* 1990; **31**(7): 987–997
304   Volek J.S., Duncan N.D., Mazetti S. *et al*. Performance and muscle fiber adaptations to creatine supplementation and heavy resistance training. *Med Sci Sports Exerc* 1999; **31**(8): 1147–1156
305   Poortmans J.R., Francaux M. Long term creatine supplementation does not impair renal function in healthy athletes. *Med Sci Sports Exerc* 1999; **31**(8): 1108–1110
306   Spector S.A., Jackman M.R., Sabounjian L.A. *et al*. Effect of choline supplementation on fatigue in trained cyclists. *Med Sci Sports Exerc* 1995; **27**: 668–673
307   Fogelholm G.M., Mäveri H.K., Kai T.K. *et al*. Low-dose amino acid supplementation: no effects on serum human growth hormone and insulin in male weight lifters. *Int J Sport Nutr* 1993; **3**: 290–297
308   Hultman E., Aderblad G., Harper P. Carnitine administration as a tool to modify energy metabolism during exercise. *Eur J Appl Physiol* 1990; **62**: 450
309   Vecchiet L., Di Lisa F., Pieralisi G., Ripari P. *et al*. Influence of L-carnitine administration on maximal physical exercise. *Eur J Appl Physiol* 1990; **61**: 486–490
310   Brazy P., Mandel J. Does the availability of inorganic phosphate regulate cellular oxidative metabolism? *N Physiol Sci* 1986; **1**: 100–103
311   Lictman M., Miller D. Erythrocyte glycolysis, adenosine triphosphate and 2,3-diphosphoglycerate concentration and adenosine hydrolysis in uremic patients: relationship to extra-cellular phosphate. *J Lab Clin Med* 1970; **76**: 267–279
312   Lictman M., Miller D., Freeman R. Erythrocyte adenosine triphosphate depletion during hypophosphatemia in a uremic patient. *N Engl J Med* 1969; **280**: 240–244
313   Cade R., Conte M., Zauner C. *et al*. Effects of phosphate loading on 2,3-diglycerate and maximal oxygen uptake. *Med Sci Sport Exerc* 1984; **16**: 263–268
314   Kreider R.B., Miller G.W., Williams M.H. *et al*. Effects of phosphate loading on oxygen uptake, ventilatory anaerobic threshold, and run performance. *Med Sci Sports Exerc* 1990; **22**: 250–255
315   Chasiotis D. Role of cyclic AMP and inorganic phosphate in the regulation of muscle glycogenolysis during exercise. *Med Sci Sport Exercs*. 1988; **20**: 545–550
316   Chanutin A., Curnish R. Effect of organic and inorganic phosphate on oxygen equilibrium and human erythrocytes. *Arch Biochem Biophys* 1969; **121**: 96–102
317   Mannix E.T., Stager J.M., Harris A., Farber M.O. Oxygen delivery and cardiac output during exercise following oral phosphate-glucose. *Med Sci Sport Exerc* 1990; **22**: 341–347
318   Kreider R.B., Miller G.T.W., Schenck D. *et al*. Effects of phosphate loading on metabolic and myocardial responses to maximal and endurance exercise. *Int J Sport Nutr* 1992; **2**: 20–47
319   Bredle D., Stager J., Brechue W., Farber M. Phosphate supplementation, cardiovascular function, and exercise performance in humans. *J Appl Physiol* 1988; **65**: 1821–1826
320   Farber M., Carlone S., Palange P. *et al*. Effect of inorganic phosphate in hypoxemic chronic obstructive lung disease patients during exercise. *Chest* 1987; **92**: 310–312

321 Farber M., Sullivan T., Finberg N. *et al*. Effects of decreased O2 affinity of hemoglobulin on work performance during exercise in healthy humans. *J Lab Clin Med* 1984; **104**: 166–175

322 Hommen N., Cade R., Privette M., Dippy J. Effect of PO4 and various fluid replacement regimens on blood volume (BV), cardiac output (Qc) and endurance during bicycle exercise. *Southeast Am C Sport Med Proc* 1953; **15**: 577–578

323 Stewart I., McNaughten L., Davies P., Tristam S. Phosphate loading and the effects on VO2max in trained cyclists. *Res Quarterly* 1990; **61**: 80–84

324 Duffy D., Conlee R. Effects of phosphate loading on leg power and high intensity treadmill exercise. *Med Sci Sport Exerc* 1986; **18**: 674–677

325 Johnson W., Black D. Comparison of effects of certain blood alkalinizers and glucose upon competitive endurance performance. *J Appl Physiol* 1953; **5**: 577–578

326 Powers S., Dodd S. Caffeine and endurance performance. *Sports Med* 1985; **2**: 165–174

327 Graham T., McLean C. Gender differences in the metabolic response to caffeine. In Tarnopolski M. (ed.). *Gender Differences in Metabolism*. CRC Press, Boca Raton, 1999, pp 301–329

328 Nehlig A., Debry G. Caffeine and sports activity: a review. *Int J Sports Med* 1994; **15**: 215–223

329 Spriet L.L. Caffeine and performance. *Int J Sports Nutr* 1995; **5**: S84–S99

330 Dodd S.L., Herb R.A., Powers S.K. Caffeine and exercise performance: an update. *Sports Med* 1993; **15(1)**: 14–23

331 Dews P.B. Caffeine. *Ann Rev Nutr* 1982; **2**: 323–341

332 Brouns F. The effect of athletic training and dietary factors on the modulation of muscle glycogen. In: Reilly T., Orme M. (eds). *The Clinical Pharmacology of Sport and Exercise*. Elsevier, Amsterdam, 1997, pp 181–197

333 Berglund B., Hemmingsson P. Effects of caffeine ingestion on exercise performance at low and high altitudes in cross-country skiing. *Int J. Sports Med* 1982; **3**: 234–236

334 Melia P., Pipe A., Greenberg L. The use of anabolic-androgenic steroids by Canadian students. *Clinical J Sports Med* 1996; **6**: 9–14.

335 Anselme F., Collomp K., Mercier B. *et al*. Caffeine increases maximal anaerobic power and blood lactate concentration. *Eur J Appl Physiol* 1992; **65**: 188–191

336 Collomp K., Ahmaidi S., Audran M. *et al*. Effects of caffeine ingestion on performance and anaerobic metabolism during the Wingate test. *Int J Sports Med* 1991; **12(5)**: 439–443

337 Graham T.E. The possible actions of methylxanthines on various tissues. In: Reilly T., Orme M. (eds). *The Clinical Pharmacology of Sport and Exercise*. Amsterdam, Elsevier, 1997, pp 257–270

338 Collomp K., Callaud A., Uadran M. *et al*. Influence du prise aigue ou chronique de caffeine sur la performance et les catecholamines au cours d'un exercise maximal. *CR Soc Biol* 1990; **184**: 87

339 Greer F., Friars D., Graham T. The effect of theophylline on endurance exercise and muscle metabolism. JAP 89 2000; 1837–1844.

340 Ivy J.L., Costill D.L., Fink W.J., Lower R.W. Influence of caffeine and carbohydrate feedings on endurance performance. *Med Sci Sports* 1979; **11**: 6–11

341 Cohen B.S., Nelson A.G., Prevost G.D. *et al*. Effects of caffeine ingestion on endurance racing in heat and humidity. *Eur J Appl Physiol* 1966; **73**: 358–363.

342 Kovacs E.M., Stegen H.C.H., Brouns F. Effect of caffeinated drinks on substrate metabolism, caffeine excretion, and performance. *J Appl Physiol* 1998; **85**(2): 709–715

343 Clinton R.B., Anderson M.E., Fraser F.F. *et al.* Enhancement of 2000 m rowing performance after caffeine ingestion. *Med Sci Sports Exerc* 2000; **32**(11): 1958–1963

344 MacIntosh B.R., Wright B.M. Caffeine ingestion and performance of a 1500 meter swim. *Can J Appl Physiol* 1995; **20**: 168–177

345 Pasman W.J., van Baak M.A., Jeukendrup A.E., de Haan A. The effect of different dosages of caffeine on endurance performance time. *Int J Sports Med* 1995; **16**: 225–230

346 Spiller M.A. The chemical components of coffee. In: Spiller G.A. (ed.). *Caffeine.* CRC Press, 1998, pp 97–161.

347 Wemple R.D., Lamb D.R., McKeever K.H. Caffeine vs caffeine-free sports drinks: effects on urine production at rest and during prolonged exercise. *Int J Sports Med* 1997; **18**: 40–46

348 Wiles J.D., Bird S.R., Hopkins J., Riley M. Effect of caffeinated coffee on running speed, respiratory factors, blood lactate and perceived exertion during 1500-m treadmill running. *Br J Sp Med* 1992; **26**: 116–120

349 Sobal J., Marquart L.F. Vitamin/mineral supplements use among athletes. *Int J. Sport Nutr* 1994; **4**: 320–334

350 Barr S.I. Nutrition knowledge of female varsity athletes and university students. *J Am Diet Assoc* 1987; **87**: 1660–1664

351 Campbell M.L., MacFayden K.L. Nutrition knowledge, beliefs and dietary practices of competitive swimmers. *Can Home Econ J* 1984; **34**: 47–51

352 Faber M., Spinnlerbenade A.J. Nutrient intake and dietary supplementation in body builders. *S Afr Med J* 1987; **72**: 831–834

353 Frederick L., Hawkins S.T. A comparison of nutrition knowledge and attitudes, dieting practices, and bone densities of postmenopausal women, female college athletes, and non-athletic women. *J Am Diet Assoc* 1992; **92**: 299–305

354 Grandjean, A.C. Profile of nutritional beliefs and practices of the elite athletes. In: Butts N.K., Gushiiken T., Zarins B. (eds.). *The Elite Athlete.* Spectrum, New York, 1985, pp 239–247

355 Grandjean A.C. Vitamins, diet and the athlete. *Clin Sports Med* 1983; **2**: 105–114

356 Graves K.L., Farthing M.C., Smith S.A., Turchi J.M. Nutrition training, attitudes, knowledge, recommendations, responsibility and resource utilisation of high school coaches and trainers. *J Am Diet Assoc* 1991; **91**: 321–324

357 Parr R.B., Porter M.A., Hodgson S.C. Nutrition knowledge and practices of coaches, trainers, and athletes. *Phys Sportsmed* 1984; **12**: 127–138

358 Short S.H., Short W.R. Four year study of university athletes' dietary intake. *J Am Diet Assoc* 1983; **82**: 632–645

359 Soper J., Carpenter R.A., Shannon B.M. Nutrition knowledge of aerobic dance instructors. *J Nutr Educ* 1992; **24**: 59–66

360 Werblow J.T., Fox H.M., Henneman A. Nutritional knowledge, attitudes and food patterns of women athletes. *J Am Diet Assoc* 1978; **73**: 242–245

361 Worme J.D., Doubt T.J., Singh A. *et al.* Dietary patterns, gastrointestinal complaints, and nutrition knowledge of recreational athletes. *J Am Diet Assoc* 1990; **51**: 690–697

362 Brill J.B., Keane M.W. Supplementation patterns of competitive male and female body builders. *Int J Sport Nutr* 1994; **4**: 398–412

363 Kleiner S.M., Bazzarre T.L., Litchford M.D. Metabolic profiles, diet, and health practices of championship male and female bodybuilders. *J Am Diet Assoc* 1990; **90**: 962–967

364 Walberg-Rankin J., Eckstein Edmonds C., Gwazdauskas F.C. Diet and weight changes of female bodybuilders before and after competition. *Int J Sport Nutr* 1993; **3**: 87–102

365 Brooks-Gunn J., Warren M.P., Hamilton L.H. The relation of eating problems and amenorrhea in ballet dancers. *Med Sci Sports Exerc* 1987; **19**: 41–44

366 Davis C., Cowles M. A comparison of weight and diet concerns and personality factors among female athletes and non-athletes. *J Psychom Res* 1989; **33**(5): 527–536

367 Pasman L., Thompson J.K. Body image and eating disturbance in obligatory runners, obligatory weightlifters and sedentary individuals. *Int J Eating Disorders* 1988; **7**(6): 759–769

368 Rosen L.W., McKeag D.B., Hough D.O., Curley V. Pathogenic weight control behaviour in female athletes. *Phys Sportsmed* 1986; **14**(1): 79–95

369 Rucinski A. Relationship of body image and dietary intake of competitive ice-skaters. *J Am Diet Assoc* 1989; **89**: 98–100

370 Sundgot-Borgen J., Corbin C.B. Eating disorders among female athletes. *Phys Sportsmed* 1987; **15**(2): 89–95

371 Walberg J.L., Johnston C.S. Menstrual function and eating behaviour in female recreational weight lifters and competitive body builders. *Med Sci Sports Exerc* 1991; **23**: 30–36

372 Evers C.C. Dietary intake and symptoms of anorexia nervosa in female university dancers. *J Am Diet Assoc* 1987; **87**: 66–68

373 Wilmore J.H. Eating and weight disorders in female athletes. *Int J Sport Nutr* 1991; **1**: 104–117

374 Sundgot-Borgen J. Prevalence of eating disorders in elite female athletes. *Int J Sport Nutr* 1993; **3**: 29–40

375 Campbell Sandri S. On dancers and diet. *Int J Sport Nutr* 1993; **3**: 334–342

376 Gordon S. *Off Balance: The Real World of Ballet.* McGraw Hill, New York, 1981

377 Gordon S. The demands of dance training. In: Peterson D., Lapenskie G., Taylor A.W. (eds). *The Medical Aspects of Dance.* Sports Dynamics, London, Ontario, 1986, pp 5–13

378 Hamilton L.H., Brooks-Gunn J., Warren M.P. Socio-cultural influences on eating disorders in professional female ballet dancers. *Int J Eating Disorders* 1985; **5**: 465–477

379 Hamilton L.H., Brooks-Gunn J., Warren M.P., Hamilton W.G. The role of selectivity in the pathogenesis of eating problems in ballet dancers. *Med Sci Sports Exerc* 1988; **20**: 560–565

380 Beals K.A., Manore M.M. Prevalence and consequences of subclinical eating disorders in female athletes. *Int J Sport Nutr* 1994; **4**: 175–195

381 Brownell K.D., Steen S.N., Wilmore J.H. Weight regulation practices in athletes: analysis of metabolic and health effects. *Med Sci Sports Exerc* 1987; **19**: 546–556

382 Horswill G.A. Weight loss and weight cycling in amateur wrestlers: implications for performance and resting metabolic rate. *Int J Sport Nutr* 1993; **3**: 245–260

383 Lakin J.A., Steen S.N., Opplinger R.A. Eating behaviours, weight loss methods and nutritional practices among school wrestlers. *J Comm Health Nurs* 1990; **7**: 59–67

384   Steen S.N., Brownell K.D. Patterns of weight loss and regain in wrestlers: has the tradition changed? *Med Sci Sports Exerc* 1990; **22**: 762–768

385   Tipton CM., Tcheng T.K. Iowa wrestling study: weight loss in high school students. *J Am Med Assoc* 1970; **214**: 1269–1274

386   Weissinger E., Housh T.J., Johnson G.O., Evans S.A. Weight loss in high school wrestling: Wrestler and parent perceptions. *Ped Exerc Sci* 1991; **3**: 64–73

387   Clark N. How to help the athlete with bulimia: practical tips and a case study. *Int J Sport Nutr* 1993; **3**: 450–460

388   Blomstrand E. *et al.* Brached chain amino acids during exercise: effects on performance and plasma concentration of some amino acids. *Eur J Physiol* 1991; **63**: 83–88

389   Van Hall G. *et al.* Ingestion of brached chain amino acids and tryptophan during sustained exercise in man: failure to affect performance. *J Physiol* 1996; **494**: 899–905

390   Blomstrand E., Hassmen P., Ek S., Ekblom B. *et al.* Influence of ingesting a solution of branched chain amino acids on perceived exertion during exercise. *Acta Physiol Scand* 1997; **159**: 41–49

391   Madsen K. *et al.* Effect of glucose, glucose plus branched chain amino acids or placebo on 100 km cycling performance. *J Appl Physiol* 1996; 2644–2650

392   Vukovich M.D. *et al.* Carnitine supplementation: effect on muscle carnitine and glycogen content during exercise. *Med Sci Sports Exerc* 1994; **26(9)**: 1122–1129

393   Trappe S. *et al.* The effects of L-carnitine supplementation and performance in man. *Int J Sports Med* 1994; **15(4)**: 181–185

394   Lambert M. *et al.* Failure of commercial oral aminoacid supplements to increase serum growth hormone secretions in male body builders. *Int Sports Nutr* 1993; **3**: 298–305

395   Anderson R.A., Brantner J.H., Polansky M.M. An improved assay for biologically active chromium. *J Agric Food Chem* 1978; **26**: 1219–1221

396   Evans G.W., Bowman T.D. Chromium picolinate increases membrane fluidity and rate of insulin internalisation. *J Inorgan Biochem* 1992; **46**: 243–250

397   Evans G.W., Pouchnik D.K. Composition and biological activity of chromium–pyridine carboxylate complexes. *J Inorgan Biochem* 1993; **49**: 177–187

398   Kaats G.R., Fisher J.A., Blum K., Adelman J.A. The effects of chromium picolinate supplementation on body composition in different age groups. *Am Aging Assoc Abstracts* 1991; **21**: 10

399   Toepfer E.W., Mertz W., Polansky M.M., *et al.* Preparation of chromium-containing material of glucose tolerance factor activity form Brewer's yeast extracts and synthesis. *J Agric Food Chem* 1977; **25**: 162–165

400   Clarkson P.M. Nutritional ergogenic aids: chromium, exercise and muscle mass. *Int. J. Sport Nutr* 1991; **1**: 289–293

401   Moore R.J., Friedl K.E. Physiology of nutritional supplements: chromium picolinate and vanadyl sulfate. *Nat Strength Cond Assoc J* 1992; **14(3)**: 47–51

402   Whitmire D. Vitamins and minerals: a perspective in physical performance. In: Berning J.R., Steen S.N. (eds). *Sports Nutrition for the 90's*. Aspen, MD, Gaithersbury, 1991, pp 129–151

403   Lefavi R.G., Anderson R.A., Keith R.E. *et al.* Efficacy of chromium supplementation in athletes: emphasis on anabolism. *Int J Sport Nutr* 1992; **2**: 111–122

404   Clancy S.P., Clarckson P.M., Decheke M.E. *et al.* Effects of chromium picolinate supplementation on body composition, strength and urinary chromium losses in football players. *Int J Sport Nutr* 1994; **4**: 142–153

405 Anderson R.A., Bryden N.A., Polansky M.M., Thorp J.W. Effects of carbohydrate loading and underwater exercise on circulating cortisol and urinary losses of chromium and zinc. *Eur J Appl Physiol* 1991; **63**: 146–150

406 Parry Billings M. *et al.* Plasma aminoacid concentrations in the overtraining syndrome: possible effects on the immune system. *Med Sci Sports Exerc* 1992; **24**: 1353–1358

407 Castell L.M. *et al.* Does glutamine have a role in reducing infections in athletes? *Eur J Appl Physiol* 1996; **73**: 488–490

408 Pedersen B.K. and Rohde T. (1997) Exercise, glutamine and the immune system. In Pedersen B.K. (ed.). *Exercise Immunology*. Springer-Verlag, Heidelberg, 1997, pp 75–87

409 Adopo E., Peronnet F., Massicotte D. *et al.* Respective oxidation of exogenous glucose and fructose given in the same drink during exercise. *J Appl Physiol* 1994; **76**: 1014–1019

410 Wurtman R.J., Hirsch M., Growdon J.H. Lecithin consumption raises serum free choline levels. *Lancet* 1977; **ii**: 68–69

411 Blusztajn J.K. *et al.* Phosphatidylcholine as precursor of choline for acetylcholine synthesis. *J Neurol Transm Suppl* 1987; **24**: 247–259

412 Allworden von H.N. *et al.* The influence of lecithin on plasma choline concentrations in triathletes and adolescent runners during exercise. *Eur J Appl Physiol* 1993; **67**: 87–91

413 Wurtman R.J., Lewis M.C. Exercise, plasma composition and neurotransmission. In: Brouns F. (ed.). *Advances in Nutrition and Topsport. Med Sportsci* Vol 32. Karger, Basel, 1991, pp 94–109

414 Zeisel S.H., Growdon J.H., Wurtman R.J. Normal plasma choline response to ingested lecithin. *Neurology* 1980; **30**: 1226–1229

415 Blusztajn J.K., Wurtman R.J. Choline and cholinergic neurons. *Science* 1983; **221**: 614–620

416 Maire J.-C., Wurtman R.J. Effects of electrical stimulation and choline availability on release and contents of acetylcholine and choline in superfused slices from rat striatum. *J Physiologie* 1985; **80**: 189–195

417 Bierkamper G.G., Goldberg A.M. Release of acetylcholine from the vascular perfused rat phrenic nerve-hemidiaphragm. *Brain Res* 1980; **202**: 234

418 Freeman J. Plasma choline: its turnover and exchange with brain choline. *J Neurochem* 1975; **24**: 729–734

419 Tucek S. *Acetylcholine Synthesis in Neurons*. Chapman & Hall, London, 1978.

420 MacIntosh F.C., Collier, B. *The Neurochemistry of Cholinergic Terminals. Handbook of Experimental Pharmocology of Neuromuscular Junction*. Springer Verlag, Berlin, 1976, pp 99–228

421 Wurtman R.J. Alzheimer's disease. *Sci American* 1985; **252**: 62–75

422 Xia N. *Effects of Dietary Choline Levels on Human Muscle Function*. Boston University College of Engineering, 1991

423 Krnjevic K., Miledi R. Presynaptic failure of neuromuscular propagation in rats. *J Physiol* 1959; **149**: 1–22

424 Pagala M.K.D., Namba T. Failure of neuromuscular transmission and contractility during muscle fatigue. *Muscle Nerve* 1984; **7**: 454–464

425 Liley A.W., North K.A.K. An electrical investigation of effects of repetitive stimulation on mammalian neuromuscular junction. *J Neurophysiol* 1952; **16**: 509–527

426 Conlay L., Wurtman R.J., Blusztajn J.K. *et al.* Marathon running decreases plasma choline concentration. *N Engl J Med* 1986; **315**: 892

427   Sandage B.W., Sabounjian L., White R., Wurtman R.J. Choline citrate may
      enhance athletic performance. *Physiologist* 1992; **35**: 236
428   Harless S.J., Turbes C.C. Choline-loading: specific dietary supplementation for
      modifying neurologic and behavioral disorders in dogs and cats. *Vet
      Med/Small Anim Clin* 1982; 1223–1231
429   Hirsch M.J., Wurtman R.J. Lecithin consumption elevates acetylcholine
      concentrations in rat brain and adrenal gland. *Science* 1978; **202**: 223–235
430   Arner P.E., Kriegholm E., Engfeldt P., Bolinder J. Adrenergic regulation of
      lipolysis *in situ* at rest and during exercise. *J Clin Invest* 1990; **85**: 893–898
431   Bergström J., Hultman E., Jorfeldt L., *et al.* Effect of nicotinic acid on physical
      work capacity and on metabolism of muscle glycogen in man. *J Appl Physiol*
      1969; **26**: 170–176
432   Douglas B.R., Jansen J.B.M.J., Jong de A.J.L., Lamers L.B.H.W. Effects of
      various triglycerides on plasma cholecystokinin levels in rats. *J Nutr* 1990; **120**:
      686–690
433   Engel A.G., Rebouche C.J. Carnitine metabolism and inborn errors. *J Inherit
      Metabol Dis* 1984; **7**: 38–43
434   Fritz I.B. The metabolic consequences of the effects of carnitine on long-chain
      fatty acid oxidation. In: Gran F.C. (ed.). *Cellular Compartmentalization and
      Control of Fatty Acid Metabolism.* Academic Press, New York, pp 39–63
435   Geser C.A., Müller-Hess R., Jéquier E., Felber J.P. Oxidation rate of
      carbohydrates and lipids, measured by indirect calorimetry, after administra-
      tion of long-chain (LCT) and medium-chain triglycerides (MCT) in healthy
      subjects. *Ernährung in der Medizin* 1974; **1**: 71–72
436   Gollnick P.D., Ianuzzo C.D., King D.W. Ultrastructural and enzyme changes in
      muscles with exercise. In: Pernow B., Saltin B. (eds). *Advances in Experimental
      Medicine and Biology*, Vol 11. Plenum Press, New York/London, 1971, pp 69–86
437   Gollnick P.D., Saltin B. Significance of skeletal muscle oxidative enzyme
      enhancement with endurance training. *Clin Physiol* 1982; **2**: 1–12
438   Gollnick P.D., Saltin B. Fuel for muscular exercise: role of fat. In: *Exercise,
      Nutrition and Energy Metabolism.* Macmillan, New York, 1988, pp 71–88
439   Green H.J., Houston M.E., Thomson J.A. *et al.* Metabolic consequences of
      supramaximal arm work performed during prolonged submaximal leg work.
      J Appl Physiol 1979; **46**: 249–255
440   Helge J.W., Richter E.A., Kiens B. Interaction of training and diet on metabolism
      and endurance during exercise in man. *J Physiol* 1996; **492**: 293–306
441   Helge J.W., Kiens B. Muscle enzyme activity in man: role of substrate
      availability and training. *Am J Physiol* 1997; **272**: R1620–R1624
442   Hoppel C.L., Davis A.T. Inter-tissue relationship in the synthesis and
      distribution of carnitine. *Biochem Soc Trans* 1986; **14**: 673–674
443   Hoppeler H., Lüthi P., Claassen H. *et al.* The ultrastructure of the normal
      human skeletal muscle. A morphometric analysis on untrained men, women
      and well-trained orienteers. *Pflügers Archives* 1973; **344**: 217–232
444   Jansson E., Kaijser L. Effect of diet on the utilization of blood-borne and
      intramuscular substrates during exercise in man. *Acta Physiol Scand* 1982; **115**:
      19–30
445   Jansson E., Kaijser L. Leg citrate metabolism at rest and during exercise in
      relation to diet and substrate utilization in man. *Acta Physiol Scand* 1984; **122**:
      145–153
446   Jeukendrup A.E., Saris W.H.M., Schrauwen P. *et al.* Metabolic availability of
      medium-chain triglycerides coingested with carbohydrates during prolonged
      exercise. *J Appl Physiol* 1995; **79**: 756–762

447   Jeukendrup A.E., Saris W.H.M., Diesen van R. *et al.* Effect of endogenous carbohydrate availability on oral medium-chain triglyceride oxidation during prolonged exercise. *J Appl Physiol* 1996; **80**: 949–954

448   Jeukendrup A., Saris W., Brouns F. *et al.* Effects of carbohydrate (CHO) and fat supplementation on CHO metabolism during prolonged exercise. *Metabolism* 1996; **45**: 915–921

449   Jeukendrup A.E., Thielen J.J.H.C., Wagenmakers A.J.M. *et al.* Effect of MCT and carbohydrate ingestion on substrate utilization and cycling performance. *Am J Clin Nutr*, 1998; **67**: 397–404.

450   Johannessen A., Hagen C., Galbo H. Prolactin, growth hormone, thyrotropin, 3,5,3' triiodothyronine, and thyroxine responses to exercise after fat- and carbohydrate-enriched diet. *J Clin Endocrinol Metabol* 1981; **52**: 56–61

451   Kiens B., Éssen-Gustavsson B., Christensen N.J., Saltin B. Skeletal muscle substrate utilization during submaximal exercise in man: effect of endurance training. *J Physiol* 1993; **469**: 459–478

452   Kiens B., Kristiansen S., Jensen P. *et al.* Membrane associated fatty acid binding protein (FABPpm) in human skeletal muscle is increased by endurance training. *Biochem Biophys Res Comm* 1997; **231**: 463–465

453   Lambert E.V., Speechly D.P., Dennis S.C., Noakes T.D. Enhanced endurance in trained cyclists during moderate intensity exercise following 2 weeks' adaptation to a high fat diet. *Eur J Appl Physiol* 1994; **69**: 287–293

454   Leddy J., Horvath P., Rowland J., Pendergast D. Effect of a high or a low fat diet on cardiovascular risk factors in male and female runners. *Med Sci Sports Exerc* 1997; **29**: 17–25

455   Massicotte D., Péronnet F., Brisson G.R. Oxidation of exogenous medium-chain free fatty acids during prolonged exercise: comparison with glucose. *J Appl Physiol* 1992; **73**: 1334–1339

456   Miller W.C., Bryce G.R., Conlee R.K. Adaptations to a high-fat diet that increase exercise endurance in male rats. *J Appl Physiol* 1984; **56**: 78–83

457   Morgan T.E., Cobb L.A., Short F.A. *et al.* Effects of long-term exercise on human muscle mitochondria. In: Pernow B., Saltin B. (eds). *Advances in Experimental Medicine and Biology*, Vol 11. Plenum Press, New York/London, 1971, pp 87–95

458   Muoio D.M., Leddy J.J., Horvath P.J. *et al.* Effect of dietary fat on metabolic adjustments to maximal VO2 and endurance in runners. *Med Sci Sports Exerc* 1994; **26**: 81–88

459   Phinney S.D., Bistrian B.R., Evans W.J. *et al.* The human metabolic response to chronic ketosis without caloric restriction: preservation of submaximal exercise capability with reduced carbohydrate oxidation. *Metabolism* 1983; **32**: 769–777

460   Pratt C.A. Lipoprotein lipase and triglyceride in skeletal and cardiac muscles of rats fed lard or glycose. *Nutr Res* 1989; **9**: 47–55

461   Romijn J.A., Coyle E.F., Sidossis L.S. *et al.* Regulation of endogenous fat and carbohydrate metabolism in relation to exercise intensity and duration. *Am J Physiol* 1993; **265**: E380–E391.

462   Soop M., Björkman O., Cederblad G., Hagenfeldt L. Influence of carnitine supplementation on muscle substrate and carnitine metabolism during exercise. *J Appl Physiol* 1988; **64**: 2394–2399.

463   Van der Vusse G.J., Reneman R.S. Lipid metabolism in muscle. In: Rowell L.B., Shepherd J.T. (eds). *Handbook of Physiology, Section 12: Exercise: Regulation and Integration of Multiple Systems.* Oxford University Press, New York, 1996, pp 952–994

464 Van Zyl C.G., Lambert E.V., Hawley J.A. *et al.* Effects of medium-chain triglyceride ingestion on carbohydrate metabolism and cycling performance. *J Appl Physiol*, 1996; **80**(6): 2217–2225

465 Lambert E.V., Goedecke J., Zyl C., Murphy K., Hawley J.A., Noakes T.D., Dennis S.C. High fat diet versus habitual diet prior to carbohydrate loading: effect of exercise metabolism and cycling performance. *Int J Sport Nutr Exerc Metab* 2001; **11**(2): 209–225

466 Vukovich M.D., Costill D.L., Hickey M.S. *et al* Effect of fat emulsion and fat feeding on muscle glycogen utilization during cycle exercise. *J Appl Physiol* 1993; **75**: 1513–1518

467 Wahrenberg H., Engfeldt P., Bolinder J., Arner P. Acute adaptation in adrenergic control of lipolysis during physical exercise in humans. *Am J Physiol* 1987; **253**: E383–E390

468 Björkman O. Fuel utilization during exercise. In: Benzi G., Packer L., Siliprandi N. (ed). *Biochemical Aspects of Physical Exercise*. Elsevier, Amsterdam, 1986, pp 245–260

469 Newsholme E.A. Basic aspects of metabolic regulation and their application to provision of energy in exercise. In: Poortmans J.R. (ed.). *Principles of Exercise Biochemistry*. *Med Sport Sci*. Karger, Basel, 1988, pp 40–77

470 Newsholme E.A. Application of knowledge of metabolic integration to the problem of metabolic limitations in sprints, middle distance and marathon running. In: Poortmans J.R. (ed.). *Principles of Exercise Biochemistry*. *Med Sports Sci*. Karger, Basel, 1988, pp 194–211

471 Beckers E.J., Jeukendrup A.E., Brouns F. *et al.* Gastric emptying of carbohydrate–medium chain triglyceride suspensions at rest. *Int J Sports Med* 1992; **13**: 581–584

472 Janssen G.M.E., Scholte H.R., Vaaandrager-Verduin M.H.M., Ross J.D. Muscle carnitine level in endurance training and running a marathon. *Int J Sports Med* 1989; **10**: S153–S155

473 Vukovich M.D., Costill D.L., Fink W.J. L-Carnitine supplementation: effect on muscle carnitine content and glycogen utilization during exercise (abstract). *Med Sci Sports and Exerc* 1994; **26**: S8

474 Maassen N., Schröder P., Schneider G. Carnitine does not enhance maximum oxygen uptake and does not increase performance in endurance exercise in the range of one hour (abstract). *Int J Sports Med* 1995; **15**: 375

475 Trappe S.W., Costill D.L., Goodpaster B. *et al.* The effects of L-carnitine supplementation on performance during interval swimming. *Int J Sports Med* 1994; **15**: 181–185

476 Jeukendrup A., Brouns F. Nutrition for endurance sports: from theory to practice. In: Jeukendrup A., Brouns M., Brouns F. *Advances in Training and Nutrition for Endurance Sports*. Isostar Sports Nutrition Foundation Conference Proceedings

477 Freeman J.J., Jenden D.J. The source of choline for acetyl choline synthesis in brain. *Life Sci* 1976; **19**(7): 949–961

478 Heinonen O.J. Carnitine and physical exercise. *Sports Med* 1996; Aug 22(2): 109–132

479 Singh A., Moses F.M., Deuster P.A. Chronic multivitamin-mineral supplementation does not enhance physical performance. *Med Sci Sports Exercise* 1992; **24**(6): 726–732

480 Weight I.M., Myburgh K.H., Noakes T.D. Vitamin and mineral supplementation: effect on the running performance of trained athletes. *Am J Clin Nutr* 1988; **47**: 192–195

481   American College of Sports Medicine. Roundtable: The physiological and health effects of oral creatine supplementation. *Med Sci Sports Exercise* 1999; **32**(93): 707–717

482   Davis J.M., Welsh R.S., De Volve K.L., Alderson N.A. Effects of branched chain amino acids and carbohydrate on fatigue during intermittent high intensity running. *Int J Sports Med* 1999; **20**: 309–314

483   Weller E., Bachert P., Meinck H.M. *et al*. Lack of effect of oral magnesium supplementation on Mg in serum blood cells and calf muscle. *Med Sci Sports Exercise* 1997; **30**(11): 1584–1591

484   Burke L.M., Hawley J.A., Angus D.J. *et al* Adaptions to short-term high fat diet persist during exercise despite high carbohydrate availability. *Med Sci Sports Exercise* 2002; **34**(1): 83–91

485   Hoppeler H., Billeter R., Horvath P.J., Leddy J.J., Pendergast D.R. Muscle structure with low and high fat diets in well trained male runners. *Int J Sports Med* 1999; **20**: 522–526

486   Whitley H.E., Humphreys S.M., Campbell I.T. *et al*. Metabolic and performance responses during endurance exercise after high-fat and high-carbohydrate meals. *J Appl Physiol* 1998; **85**(2): 418–424

487   Burke L.M., Claassen A., Hawley J.A., Noakes T.D. Carbohydrate intake during prolonged cycling minimizes effect of glycemic index of preexercise meal. *J Appl Physiol* 1998; **85**(6): 2220–2226

488   Balsom P.D., Wood K., Olsson P., Ekblom B. Carbohydrate intake and multiple sprint sports: with special reference to football (soccer) *Int J Sports Med* 1999; **20**: 48–52

489   Tsintzas K. and Williams C. Human muscle glycogen metabolism during exercise: effect of carbohydrate supplementation. *Sports Med* 1998; Jan 25(1): 7–23

490   Pitsiladis Y.P., Smith I., Maughan R.J. Increased fat availability enhances the capacity of trained individual to perform prolonged exercise. *Med Sci Sports Exercise* 1999; **31**(11): 1570–1579

491   Gisolfi C.V., Summers R.W., Lambert G.P., Xia T. Effect of beverage osmobility on intestinal fluid absorption during exercise. *J Appl Physiol* 1998; **85**(5): 1941–1948

492   Jeukendrup A.E., Wagenmakers A.J.M., Stegen J.H. *et al*. Carbohydrate ingestion can completely suppress endogenous glucose production during exercise. *Am J Physiol* 1999; **276** (endocrinol metab 39) E672–E683.

493   Graham T. Caffeine and Coffee: A useful supplement? Insider News on Sports Nutrition, Isostar Sport Nutrition Foundation. 1998; **6**(2): 1–8

Index

Lightning Source UK Ltd.
Milton Keynes UK
10 July 2010

156774UK00001B/66/P